SpringerBriefs in Earth Sciences

For further volumes:
http://www.springer.com/series/8897

Heriberta Castaños · Cinna Lomnitz

Earthquake Disasters in Latin America

A Holistic Approach

Springer

Heriberta Castaños
Instituto de Investigaciones Económicas
Universidad Nacional Autónoma de México
Mexico, DF
Mexico
e-mail: bety@unam.mx

Cinna Lomnitz
Instituto de Geofísica
Universidad Nacional Autónoma de México
Mexico, DF
Mexico
e-mail: cinna@prodigy.net.mx

ISSN 2191-5369
ISBN 978-94-007-2809-7
DOI 10.1007/978-94-007-2810-3
Springer Dordrecht Heidelberg London New York

e-ISSN 2191-5377
e-ISBN 978-94-007-2810-3

Library of Congress Control Number: 2011942349

© The Author(s) 2012
No part of this work may be reproduced, stored in a retrieval system, or transmitted in any form or by any means, electronic, mechanical, photocopying, microfilming, recording or otherwise, without written permission from the Publisher, with the exception of any material supplied specifically for the purpose of being entered and executed on a computer system, for exclusive use by the purchaser of the work.

Disclaimer: The facts and opinions expressed in this work are those of the authors and not necessarily those of the publisher.
Every effort has been made to contact the copyright holders of the figures and tables which have been reproduced from other sources. Anyone who has not been properly credited is requested to contact the publishers, so that due acknowledgment may be made in subsequent editions.

Printed on acid-free paper

Springer is part of Springer Science+Business Media (www.springer.com)

Preface

> *Forsake, dear friend, all withered theories*
> *And hug the greening, golden Tree of Life.*
>
> Faust

Charles Darwin's discoveries in the natural sciences began during his South America cruise. On February 20, 1835, during the Beagle's voyage around South America, he felt and described the great Concepción, Chile megaquake. In September of the same year he described the selective evolution of birds on the Galapagos Islands. The latter discovery eventually led him to formulate the Theory of Evolution.

The fate of Darwin's observations on earthquakes is less well known. Darwin intriguingly described seismic waves of the Chile megaquake as feeling like ocean waves, though he was standing on solid ground. His description was not taken at face value, partly because later work by Lord Rayleigh seemed to show that surface waves on land and at sea were fundamentally different. But Darwin was a shrewd observer and he may have been right all along.

This book is an attempt to demonstrate the analytical power of a holistic approach to the understanding of disasters. We take six major earthquakes in Latin America as an example of unity in diversity. Understanding disasters is a way of understanding the social system. The idea is showing that every major disaster is unique and different. We start out with discussing plate seismology and some basics of earthquake engineering. Why was Mexico City founded in a lake? Why is earthquake damage rising exponentially everywhere and why is this trend unsustainable? Why is history relevant for understanding earthquake risk?

The purpose of this SpringerBrief volume is to discuss the causes of severe damage in recent megaquakes in Latin America and how to prevent the recurrence of human and material losses caused by earthquake disasters. Future SpringerBriefs may apply a holistic approach to other types of disasters and in other regions of the world.

No two earthquakes are identical. Even if they were, the society is different. Building codes are attempts to adapt engineering experience to local conditions: they need to be updated. And so do theories. Every disaster teaches the same lesson in a different way.

The United States is spending an average of one billion dollars a week in prevention and mitigation of natural disasters, including earthquakes, floods, hurricanes, winter storms, and others. Latin America has its share of megaquakes: its societies need to spend more on disaster prevention and disaster research. We appreciate the valuable assistance of our students, Luz Aída Lozano Campos and Miguel Ángel Aguilar Dorado.

Contents

1 Darwin and Plate Tectonics 1
  1.1 Darwin Feels an Earthquake 1
  1.2 Earthquake Physics 4
  1.3 Plate Tectonics and Failure Modes 5
  1.4 How to Measure Earthquakes 6
  1.5 Failure Modes 7
  1.6 Tsunami Warning Systems 8
  1.7 Megaquakes and Transelastic Waves 12
  References 13

2 The Great 1960 Chile Megaquake 15
  2.1 A Great Earthquake 15
  2.2 Earthquakes and Coastal Geomorphology 17
  2.3 An Observation of Long-Period Surface Waves
      on Soft Ground 19
  References 20

3 The 1967 Caracas Earthquake 21
  3.1 Caracas in 1967 21
  3.2 Accountability and the Law 23
  References 24

4 The 1970 Peru Earthquake 25
  4.1 The Santa Valley 25
  4.2 Geology of the Disaster 27
  4.3 Climate Change and Communicative Action 28
  References 30

5 The 1985 Mexico Earthquake ... 31
5.1 September 19, 1985, 7:19 a.m. ... 31
5.2 Lázaro Cárdenas ... 33
5.3 Mexico City ... 35
5.4 Social Effects of the Earthquake ... 36
5.5 The Senate Hearings ... 37
5.6 Disaster, Denial, and the Law ... 38
References ... 39

6 The 2010 Haiti Earthquake ... 41
6.1 An Unmitigated Catastrophe ... 41
6.2 Possible Causes of the Disaster ... 42
6.3 Disaster Culture in an Earthquake Country ... 43
Reference ... 45

7 The 2010 Chile Earthquake ... 47
7.1 Another Megaquake ... 47
7.2 Earthquake Damage Caused by Long-Period Surface Waves ... 49
7.3 Some Historical Sidelights ... 49
7.4 Disaster and Regional Development ... 51
Reference ... 53

8 A List of Significant Earthquakes in Latin America ... 55
8.1 Introduction ... 55

9 Conclusions and Recommendations ... 57
9.1 Summary ... 57
9.2 Megaquakes ... 58
9.3 Sustainability ... 58
9.4 Tsunamis ... 59
9.5 Disaster Culture ... 60
9.6 Building Codes ... 61
9.7 Engineering Ethics ... 62
  9.7.1 Six Commandments of the Engineer ... 63
9.8 Recommendations ... 64
References ... 65

Chapter 1
Darwin and Plate Tectonics

Abstract Charles Darwin felt and described the great 1835 Chile earthquake which destroyed Concepción and raised the coastline. He was the first scientist to suggest that earthquake waves could be similar to ocean waves. He also thought earthquakes might be caused by tectonic deformation of the earth. The physics of earthquakes is introduced and it is shown how prograde ground motion might destabilize buildings. Tsunami warning systems are critically examined.

1.1 Darwin Feels an Earthquake

His Majesty's Ship *Beagle*, a ten-gun, two-mast, square-rigged brig, sailed from Plymouth Sound two days after Christmas of 1831, or 4 years before this story begins (Fig. 1.1). It was late summer in the Southern hemisphere and the Beagle was strenuously wending her way northward along a little-explored stretch off the Pacific coast of South America. The frail vessel encountered unseasonably poor weather. Captain Robert FitzRoy, the moody and aristocratic commander of the *Beagle,* guided the small craft beating to windward. They finally reached the small Chilean outpost of Valdivia on the night of February 8, 1835.

During the following days, FitzRoy busied himself with hydrographic measurements while the young naturalist on board, Mr Darwin, spent his 26th birthday exploring the rain forests with a local guide. On February 20, shortly before noon, he was resting in a forest clearing when a major earthquake occurred. The epicenter was some 300 km to the north. Charles Darwin describes the experience as follows.

> It came on suddenly, and lasted two minutes, but the time appeared much longer. The rocking of the ground was very sensible There was no difficulty in standing upright, but the motion made me almost giddy: it was something like the movement of a vessel in a

Fig. 1.1 H.M.S. *Beagle* at anchor off the coast of Patagonia (after a contemporary drawing)

little cross-ripple, or still more like that felt by a person skating over thin ice, which bends under the weight of his body. A bad earthquake at once destroys our oldest associations: the earth, the very emblem of solidity, has moved beneath our feet like a thin crust over a fluid (Darwin 1845)

Darwin's reference to giddiness was significant. He suffered from seasickness: years on board the *Beagle* had done nothing to overcome this infirmity. Seasickness is a kind of motion sickness or kinetosis which affects the vestibular system of the inner ear. Three semi-circular canals in each ear contain a fluid that moves against tiny sensors to detect rotations of the head. This system will react to some types of motion by inducing a feeling of vertigo or giddiness.

In one short paragraph, Darwin made short shrift of age-old prejudice. He made four separate references to fluid motion, yet he knew he was on solid ground. But Valdivia was located inside a deep bay where three major rivers converged. The main streets of Valdivia were canals, as in Venice. The settlement was on soft, partly swampy terrain. And soft soils are recognized as intermediate materials between solids and liquids. The state transition between solid and fluid soil has been explored chiefly by soil mechanics. It is not as clear-cut as it would be in water. Before a granular material like sand or clay can flow like a liquid it must first traverse a borderline state known as a *mesophase*. During the mesophase and before reaching the point of liquefaction the material is a solid: it supported Darwin's weight. But the intergranular cohesion dropped to a point where it would no longer propagate elastic waves. Instead, gravity took over as the restoring force.

As the *Beagle* sailed into the harbor of Concepción on March 4th 1835, the British explorers were confronted with a dire spectacle. They were amazed to find the town totally destroyed by the earthquake. The port area had been washed away by huge tsunami waves and there were many dead in town.

It had not been the first megaquake to afflict the locality. Concepción is the second-largest city of Chile today, but it continues to suffer severe damage from successive megaquakes. The epicenter of the 1835 megaquake was probably just off Concepción. Robert FitzRoy and Charles Darwin explored the surroundings and discovered that the offshore island of Santa María had been raised bodily out of the water. The shallow sea bottom around the island had emerged and become visible at low tide. Beds of dead mussels were still clinging to the rock, ten feet above the high-water mark. Darwin concluded that the 1835 Concepción earthquake had been caused by uplift of the offshore islands. He boldly suggested that *all* earthquakes might be caused by tectonic movements:

> The most remarkable effect of this earthquake was the permanent elevation of the land; it would probably be far more correct to speak of it as the cause (Darwin 1845).

The measurements suggested that the Bay of Concepción had been raised "by two or three feet". This figure is consistent with observed uplift in the 2010 megaquake. But Darwin went beyond: if it was true that mountains and earthquakes had the same cause, the uplift could imply that the earth is considerably older than what scientists believed in 1835. When Darwin explored the hills around Valparaiso he found sea shells at an elevation of 1,300 feet. If the coast of Chile rose at the rate of three feet per century, it should have taken earthquakes more than four hundred centuries to lift the sea shells to their present elevation. Yet calculations by the best theologians yielded an age of only fifty centuries for the time of Creation.

Darwin was religious, and this train of thought disturbed him deeply. For a time he kept his discovery to himself. But what sins, he mused, could justify the suffering of the citizens of Concepción? The city was located on a river terrace of soft sediments fronting on the Bio–Bio River. Liquefaction was observed on the river banks after the 1835, 1960 and 2010 events. Darwin describes the harrowing experience of a survivor:

> Mr Rouse, the English consul, told us that he was at breakfast when the first movement warned him to run out. He had scarcely reached the middle of the courtyard when one side of his house came thundering down. He retained presence of mind to remember that if he once got on the top of that part which had already fallen, he would be safe. Not being able from the motion of the ground to stand, he crawled up on his hands and knees; and no sooner had he ascended this little eminence than the other side of the house fell in, the great beams sweeping close in front of his head. With his eyes blinded and his mouth choked with the cloud of dust which darkened the sky, at last he gained the street... The thatched roofs fell over the fires, and flames burst forth in all parts. Hundreds knew themselves ruined, and few had the means of providing food for the day. (Darwin 1845).

Darwin slightly misinterpreted his observations during the earthquake. He thought the feeling of dizziness was caused by some unknown perturbation of his senses: "*A bad earthquake at once destroys our oldest associations.*" He should have trusted his sensory organs. Actually the feeling of being on a liquid was correct. Strong earth vibrations had brought the soil closer to liquefaction, so that the surface no longer described retrograde elliptical motion as on a solid. Instead the soil particles moved in a prograde ellipse, just like an ocean wave. This difference was real; it was translated into giddiness by his inner ear.

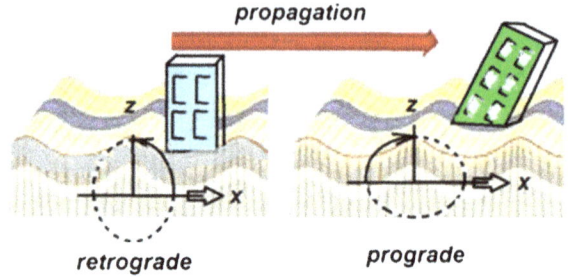

Fig. 1.2 Retrograde waves (*left*, on rock) tend to stabilize a structure but prograde waves (*right*, on soft ground) cause it to lean into the direction of propagation. The additional overturning moment may produce collapse

Earth science has made some strides since Charles Darwin visited the Latin American continent, yet his observations of the 1835 Chile megaquake (M about 8.5) remain valid and intriguing as they were 175 years ago. When Darwin visited Latin America he was a young man and Latin America was young. The continent was still largely unexplored, yet nearly everybody had lived through some natural disaster—a hurricane, a drought, most often an earthquake. When it comes to earthquakes, we have plenty of personal experiences. We cannot afford to be scientific skeptics: it would be like doubting the Devil when he holds us by the scruff of the neck.

Let us grab our doubts and throw them out. We need a better science, and more of it. Science, it is said, can have its perils: but closing our minds to truth can be infinitely more dangerous.

1.2 Earthquake Physics

Earthquakes, it is agreed, cause surface waves to spread out from the epicenter in all directions. Any point on the surface will rotate in an ellipse. It is assumed that the effect on a building is linear with peak ground acceleration (pga). But is it?

For physical reasons water waves are *prograde* while elastic surface waves are *retrograde*. Thus the sense of rotation is opposite in water and on solid ground. But the transition between both types of surface waves is highly nonlinear. So is the transition between stability and collapse. A surfer riding the crest of a water wave moves forward with the speed of the wave: but on taking a fall she is propelled first forward and upward, then downward and backward. This is prograde motion. Loss of stability is a nonlinear function of the slope of the wave.

In a Rayleigh wave the sense of rotation is exactly the opposite from water waves, namely first upward and backward, then forward and down. A building on solid rock will be propelled precisely in this fashion (Fig. 1.2). Thus a Rayleigh wave (a technical name for a surface wave on an elastic solid) helps the building preserve stability, while a water wave contributes toward causing a surfer to lose her balance. The vestibular system of the inner ear can certainly tell the difference between prograde and retrograde motion.

Charles Darwin knew this from experience. He did not know the direction of propagation: in fact, he remarked on people being confused about telling where the waves came from. But he would feel an acute discomfort when the ground motion mimicked the sense of rotation of an oceangoing vessel. He did not know the reason: in 1835 nobody did. John William Strutt, the Lord Rayleigh and discoverer of Rayleigh waves, was not born yet. He would be born 7 years after the 1835 earthquake, and it would take geophysicists even longer to find out why soft ground conditions can cause Rayleigh waves to change their sense of rotation from retrograde to prograde.

1.3 Plate Tectonics and Failure Modes

A *failure mode* is a systemic or structural weakness potentially capable of causing an engineering system to collapse. Identifying failure modes in advance is important but difficult: because of the complexity of a structure it may require considerable insight or intuition on the part of the engineer. However, this kind of foresight pays off as it may often prevent future casualties and damage.

Projective insight by itself might not guarantee that the structure will survive a major earthquake but it helps. A professional study of the performance of engineering structures in disasters is called a "failure mode and effects analysis" (FMEA). Because of the wide availability of computer programs that do the designing, engineers may be tempted to trust computers to do the job. But no computers can replace intuition aided by experience. Computer-aided design can go a long way to help a good engineer understand what an earthquake will do to his structure: but the increasing complexity of modern structures is more demanding because of the emergence of unforeseen modes of excitation.

When designing a new structure, the first thing to do is find out where we are. There are fifteen major tectonic plates plus a large number of minor plates (Fig. 1.3). Each tectonic plate is represented by a different shade. The Pacific Plate is the largest, and the Caribbean Plate is one of the smallest. The mosaic of major plates represents the position of the continents today, but because all plates are moving the configuration of continents has changed during geologic history. It is changing today: the small arrows indicate how the plates are moving. Thus the South American Plate is moving away from the African Plate and this motion causes the Atlantic Ocean to grow while the Nazca Plate is shrinking in size at a rate of 8 cm/year—the rate of subduction under the South American Plate.

This speed of subduction is relatively fast as plate motions go. Of course, it is just an average speed. Most of the time, nothing happens: the two plates are comfortably squeezing and squishing against each other as if made of putty. Then, suddenly, an earthquake happens. The boundary breaks and the plates move suddenly by ten or twenty meters. Such a high rate of subduction helps explain the high seismic hazard along the coast of the Pacific Ocean.

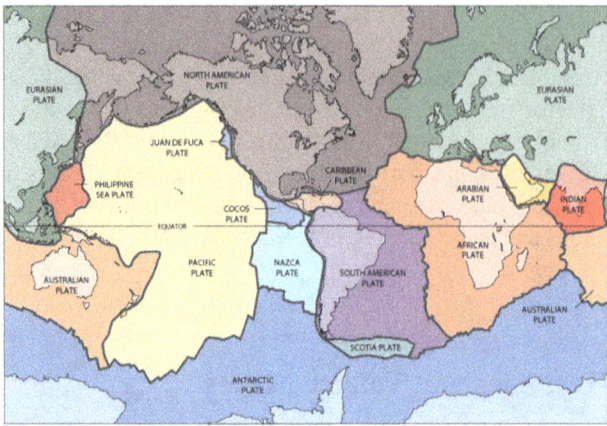

Fig. 1.3 Plate tectonics: Subduction at plate boundaries is the most common cause of damaging earthquakes. After Wikipedia Commons

1.4 How to Measure Earthquakes

Measuring earthquakes is a full-time occupation. Seismologists work routinely to assign parameters such as magnitude, epicentral location, and amount of damage to thousands of instrumental earthquakes recorded in their bailiwick. The most famous of these parameters is *magnitude*.

The magnitude M of an earthquake is computed from the logarithm of the maximum amplitude of the earthquake record; it depends on the *epicentral distance* from the epicenter to the station. Thus the first thing to do is locate the earthquake, as an estimation of magnitude is only as good as the accurate location of the event. This takes good data and it takes time. In Latin American earthquakes, epicentral locations will change as more data come in, as the early estimates are often in error by 50 km or so. This is due to relatively poor coverage by seismic stations.

Richter (1900–1985) is credited with inventing the magnitude scale. He proposed a definition of earthquake magnitude in 1935, based on an analogy with the magnitude of stars. His original idea was that M should be proportional to the logarithm of the amplitude as measured at Pasadena to the nearest integer, but as more and more people wanted to measure magnitudes his formula was modified and perfected by various scientists and now magnitudes are determined to the nearest first decimal. There are several different magnitude scales but the one most commonly used is the moment magnitude Mw.

Why not just use the energy E of earthquakes and forget about magnitude? The answer is that estimating the energy of earthquakes requires integrating seismic records over the surface of the earth. Magnitudes can be estimated from the measured amplitudes at a single location, or at a few locations. This is more practical. An old formula by Richter is often used to convert magnitude M to energy E in joules:

$$\log_{10} E = 1.5M + 4.8.$$

For example, a megaquake such as the 2011 Tôhoku, Japan earthquake is equivalent to the yield of 25,000 nuclear bombs. Such figures are beyond the grasp of an ordinary human. It is easier to use the magnitude scale and say: $M = 9.0$.

1.5 Failure Modes

Figure 1.2 shows two buildings, one on hard ground and the other on soft ground. The building on hard ground (*left*) is subjected to retrograde rotation while the one on soft ground (*right*) experiences prograde rotation. Consider what happens in both cases.

Retrograde ground motion tilts the ground away from the direction of propagation. Does this stabilize the building, or cause it to lose stability? It all depends on the way the ground tilts. All else being equal, the overturning moment (xz) is of opposite sign from the tilt dx/dz. Thus retrograde ground motion helps the building preserve stability, as it works against the earthquake forces. The effect of tilt is known by engineers as "P-Δ effect". In conclusion, a building on hard rock will resist surface waves. This does not necessarily mean that it cannot lose stability: it can, provided that the horizontal forces from the earthquake overpower the P-Δ effect. But the P-Δ effect works for stability.

What happens if the building stands on soft ground? Initially, nothing. The earthquake will not overturn the building as long as the earthquake motion is weak. However, any earthquake will tilt the surface into the direction of propagation, because dx/dz is of the same sign as *(xz)*. In a strong earthquake the resulting P-Δ effect adds to the overturning moment and increases the horizontal force of the earthquake, and this combined effect may overwhelm the stability of the structure. The building may collapse.

Fortunately not all tall buildings on soft ground will collapse in earthquakes. In the 1985 Mexico earthquake 16% of them did. However, this is bad news. A simple reasoning explains why soft ground is more hazardous than hard ground in an earthquake. Soft ground conditions are among the most common causes of loss of life in Latin American earthquake disasters. To put it briefly: mud can be dangerous.

A very soft soil will behave much like water because most of it is water. In strong earthquakes the shear resistance of these materials will decay until it is extremely low. Water has a very low shear resistance: not zero, as one might suppose, but very low. Mud is an in-between material: not a rock, not a fluid, just in-between. How right was Darwin when he compared soft ground with water during an earthquake! But the mechanical properties of mud are still largely unexplored.

Malischewsky et al. (2008) calculated the bounds of existence of Rayleigh waves in very soft materials. At low amplitudes the motion is retrograde but prograde wave motion begins to dominate as Poisson's ratio approaches $v \to 0.5$, which is the case in soft saturated soils. Poisson's ratio is a function of the ratio of sound velocity to shear wave velocity in a material. In a typical rock of the earth's crust this ratio is $R \sim 1.76$. In Mexico City mud it is $R = 1500/50 = 30$ or seventeen times the value of R in a rock, because the shear velocity is Vs = 50 m/s and the velocity of sound is Vp = 1500 m/s. From R we may find Poisson's ratio v by

$$v = (3R^2 - 2)/(6R^2 + 2) = 0.499$$

which spells disaster. It is close to 0.5—too close for comfort! Water, of course, has a Poisson's ratio of $v = 0.5$; the value in hard rock should be around $v = 0.27$.

In conclusion, the behavior of soft wet sediments can be much closer to water than to rock. The effect of a mud wave impacting a building can be severe: it is like a water wave hitting a ship broadside. John A. Blume, a famous earthquake engineer, used to put it this way: design your structure on soft ground against earthquakes as if it were a boxing champion in the ring. It should "roll with the punches."

1.6 Tsunami Warning Systems

Megaquakes are very large, very infrequent and terrifyingly dangerous seismic events. They are accompanied with huge, destructive tsunamis. Most seismologists would define a megaquake as a tsunamigenic earthquake above magnitude 8 or 8.5. It makes them hard to design against. Building codes are definitely *not* designed against megaquakes. This is not lack of foresight: just plain commonsense. The average home is supposed to last 30–40 years, the average turnover of human generations. Unlike cavemen we build our dwellings to the taste of the times. But megaquakes have long average return periods, much longer than a century.

Why are these large earthquakes so difficult to predict? Latin America has its share of megaquakes and it makes sense to find out how and where they occur, what they have in common, how to predict them, and what to do about them. But they are rare events.

A typical megaquake is a large thrust earthquake that occurs in an active subduction zone and that causes a major tsunami. Megatsunamis cause more victims than does the severe shaking inland. But the tsunami wave is delayed because its propagation speed is much slower than the propagation of the earthquake wave in rock. Tsunami casualties in the near field will occur after the tsunami hits the coast, within half an hour after the earthquake. These casualties are preventable by timely withdrawal of populations from the coast.

1.6 Tsunami Warning Systems

Coastal evacuation is feasible. But here's the rub: how to predict a megaquake within, say, a week to a month in advance, and how to identify it as a megaquake within minutes—or even better, while it is still occurring? The question is of more than academic interest. It is essential to use efficiently the delay between the earthquake and the first wave of the tsunami to hit the coast, so that the endangered population can be evacuated.

Suppose that the typical delay between earthquake and tsunami is on the order of 30 min. Most people who have survived a mega-tsunami did not wait to be told that a wave was coming: they felt the earthquake and assumed that a tsunami was on the way. So they took to the hills. This is often a sensible decision. The tsunami "runup"—its maximum height above mean sea level—is 20 or 30 m in mega-tsunamis. So don't bother to get dressed for the occasion: catching a cold is not as bad as drowning in a tsunami. The alternative is most unpleasant. But let us see what actually happens.

In the 2011 Mw 9.1 Tôhoku, Japan megaquake it took the US Geological Survey (USGS) more than 38 min to estimate the true magnitude of the earthquake and identify it as a megaquake. It takes time to recognize whether a large earthquake is a megaquake or just "a large earthquake". Ordinary earthquakes are more frequent and do not generate giant tsunamis. And we don't want to ask millions of people to drop everything and run from the coast unless there is a real tsunami afoot.

The following response sequence corresponds to the 2011 Tôhoku earthquake but it is typical of what may happen in Latin America. The first data released to the public originated from the National Tsunami Warning Center in Honolulu: it was based on information supplied by the USGS and the Japan Meteorological Agency, the local agency in charge of earthquake information. It indicated that the earthquake had a moderate magnitude of 7.9. As a result, the PAGER system of the USGS at the National Earthquake Information Center (NEIC) in Golden, Colorado released a "green alert" which means that "there is a 72% chance of not getting more than *one* fatality." This initial estimate was published online and forwarded to government and critical facilities. It was misleading: the actual magnitude of the event was 9.1 (Hayes et al. 2011).

This was the state of official information in Japan right after the earthquake. It was the available information when the tsunami started to hit the coast. After 33.6 min counting from the origin time of the megaquake, the Tsunami Warning Center in Honolulu again called NEIC in Golden, Colorado and the Japan Meteorological Agency. Now they were aware that the earthquake was substantially larger and much more dangerous than a 7.9 event. They suggested preparing a new joint release to the public based on new incoming data. The magnitude now looked much higher, and the Tôhoku earthquake was recognized as a megaquake. An updated report was released to the web site of NEIC 43 min after the earthquake. The PAGER alert was upgraded from "yellow" to "red" and the estimation of fatalities was raised from green to yellow ("51% probability of getting between 10 and 100 fatalities"). Also, the estimation of economic losses was upgraded to red ("41% probability of between 1,000 and 10,000 million dollars of economic losses"). These predictions still fell short of the grim reality.

The predicted losses from PAGER had already been overtopped by the time the information was released. Worse, the information reached the population in the endangered area much too late to do anything. However, it was in time for people in more remote areas to prepare for adopting measures against an eventual tsunami wave. Therefore this performance of the warning system was described as a "dramatic improvement" over the response to the 2004 Sumatra–Andaman Islands megaquake. But it was not good enough to save people in the immediate area. The agencies were doing all they could but the data—in this case, ten automatic data channels from northern Japan—were insufficient. As a rule, less than a dozen data sets will frequently lead to an underestimation of the true size of an earthquake.

After the 2010 Chile earthquake the Chilean Navy was harshly criticized for its failure to warn the coastal population (and the country's president!) about the true size of the tsunami. The record shows that the performance of the Chilean warning system was probably as good as was the response capacity of the system as a whole. The Chilean Navy Hydrographic Service (SHOA) was in charge of tsunami surveillance and warning since 1966. It is a member of the International Tsunami Warning Center (ITIC) in Honolulu. For tsunamigenic earthquakes with epicenter off the Chilean coast, fast communication between Honolulu and SHOA in Valparaiso is of the essence. But Honolulu has no observational facilities of its own: it depends on processed seismic information from the USGS in Golden and from local agencies around the Pacific rim. Communications must be dependable and rapid, both ways.

The earthquake occurred after 3 a.m. on a summer vacation weekend. The earthquake caused a nationwide power blackout. A cautionary warning message from ITIC was received at SHOA right after the earthquake but there was no reply. Why did the Chilean Navy personnel on the midnight watch fail to reply to the cautionary message? They had strongly felt the earthquake, there was damage in town, and a warning from Honolulu to the effect that "a strong earthquake had just occurred" was superfluous. Valparaiso had been hit and people had probably been killed. It was an emergency. What SHOA wanted to know was: had Honolulu any information or advice about the tsunami? But Honolulu had quoted Golden which placed the focus of the earthquake slightly on land and at depth, adding that "no tsunami had been confirmed."

In other words, Honolulu did not know, and expected Valparaiso to *tell them* whether a tsunami had been generated. But Valparaiso was in a blackout and there were no communications with the epicentral region—or with the capital Santiago. They were in the dark. About 15 min later the tsunami was reaching the nearest points on the Chilean coast. Honolulu decided to phone Chile but unfortunately, the officer in charge spoke no Spanish and the Chileans spoke no English. The result was utter confusion. The Chileans assumed that the preliminary epicenter location and magnitude estimate were final. It appeared that the earthquake was of moderate size—not a megaquake—and that no tsunami had been generated *because the epicenter was on land and the focus was too deep.*

1.6 Tsunami Warning Systems

Fig. 1.4 The 2010 tsunami turned the coves of southern Chile into junkyards. From Wikipedia

The Hydrographic Office in Valparaiso was in trouble. The chief officer of SHOA could not be reached—cell phones were not working and he came in an hour late, because of the blackout—and there were no communications with the National Emergency Office in Santiago. This office is the sole authorized agency to alert the population. When a radio link was established it was of poor quality and there are conflicting versions of what was actually said. Did the Navy people tell Santiago "not to worry" about a tsunami? Did the Navy transmit a "red alert" to Navy stations along the coast? President Bachelet of Chile made no mention of a possible tsunami in her first broadcast message, when two major tsunami waves had impacted the coast. The waves were big as a two-story house and there were hundreds of dead.

The Navy personnel in charge of SHOA might have been aware that initial locations of earthquakes in the South American region can be 50 km off, because of sparse coverage by local earthquake stations. Afterwards the President complained about having been misinformed and the responsible officials at the Hydrographic Office and the National Emergency Office were sacked. But the available evidence suggests a communication breakdown from both ends.

This situation is far from uncommon. Earthquakes occur along the boundaries between an oceanic plate and a continental plate. The distribution of earthquake stations is lopsided—all stations are on the continental plate and none are on the oceanic plate. This introduces a bias in the locations. Eventually, relocation by USGS confirmed that the epicenter of the 2010 Chile megaquake was offshore and that the focal depth was normal for thrust earthquakes. Also, most member states of the international warning center for the Pacific rim were Spanish-speaking.

Fortunately most coastal residents and several thousand tourists had enough sense to seek shelter on nearby hills. The full moon lighted their way despite the blackout. The unlucky ones did not make it. Reportedly, some of them interpreted the absence of an alert as meaning that there was no tsunami risk (Fig. 1.4).

1.7 Megaquakes and Transelastic Waves

Up to 2011 only six large earthquakes have been recognized as megaquakes: the 1960 and 2010 Chile earthquakes (Mw = 9.5 and 8.8), the 1964 Alaska earthquake (Mw 9.2), the 1770 Cascadia, north-west coast of North America, earthquake (Mw 9.0), the 2004 Sumatra–Andaman Islands earthquake (Mw 9.3) and the 2011 Tôhoku, Japan earthquake (Mw 9.0). These earthquakes had some features in common: all were destructive, and all were tsunamigenic.

Darwin discovered that there were striking similarities between the ground motion in the 1835 megaquake and water waves. The presence of prograde surface waves could explain the high incidence of heavy damage in modern buildings on soft ground. But why is this type of ground motion not identified as a relevant feature of megaquakes? Is it possible that prograde ground motion is not being detected by seismic stations? This possibility seems remote, but perhaps not as remote as one might think.

Systematic errors in magnitude and location are not unknown. The size of very large earthquakes has been frequently underestimated, because they are rare and also because seismic stations do not provide a representative sampling of the earth's surface. Most of the earth's surface is covered by deep layers of water and soft soil. Yet most earthquake stations are located in hard rock sites, where conditions were more similar to those in the Earth's interior.

Theory predicts that the sense of rotation of surface particles is reversed when soft soils are excited to high amplitudes. This reversal occurs when Poisson's ratio tends to $v = 0.5$, the value of Poisson's ratio in water. But in soft soils the value of Poisson's ratio changes with the amplitude of the mechanical excitation. The rigidity or shear modulus μ in a soil decays inversely with the strain amplitude τ:

$$\mu = \mu_0 \frac{1}{1+\tau},$$

where μ_0 is the rigidity of the unstrained specimen and $\tau = \varepsilon/\gamma_r$ is the normalized strain amplitude. The material constant γ_r is called the "reference strain" (Hardin and Drnevich 1972).

As vibration gets stronger and the earthquake increases in duration the material weakens until it finally flows like a liquid. This transition is similar to melting, except that melting occurs at a critical temperature (the *melting point*), while liquefaction in soils occurs more gradually. There is an upper range—a mesophase—where the sense of rotation of particles is reversed. This reversal will occur well before the soil liquefies.

This peculiar property has been described as "strain softening" in soils. In water waves the restoring force is gravity, while in granular solids it is intergranular cohesion. These two forces are independent, but both will act on all kinds of materials. In soils the effect of gravity is usually neglected but both effects can be of comparable magnitude. When cohesion decays, as shown in the preceding equation, gravity may take over. At this point the surface wave turns prograde, even if the material preserves some residual cohesion.

The existence of hybrid, gravitationally perturbed Rayleigh waves was predicted by theoreticians (see, e.g., Gilbert 1967). They were not widely observed because we have not looked for them. The motion in the neighborhood of a point on the surface is fully specified by all six independent values of ε and ξ, the components of the strain tensor. But seismic stations record only the translational strain tensor ε and not the rotational tensor ξ, where

$$\varepsilon_{ik} = \frac{1}{2}\left(\frac{\partial v_k}{\partial x_i} + \frac{\partial v_i}{\partial x_k}\right), \xi_{ik} = \frac{1}{2}\left(\frac{\partial v_k}{\partial x_i} - \frac{\partial v_i}{\partial x_k}\right)$$

and x_i, v_i are the components of the rectangular coordinates and of the velocities (Jeffreys 1970). In order to observe prograde transelastic surface waves we have to look for the rotational components of ground motion. They must be monitored, especially where soft ground conditions represent a serious hazard.

References

Darwin C (1845) Journal of researches into the natural history and geology of the countries visited during the voyage of H.M.S. Beagle round the world, under the command of Capt. FitzRoy, R.N. John Murray, London
Gilbert F (1967) Gravitationally perturbed Rayleigh waves. Bull Seismol Soc Am 57:783–794
Hardin B, Drnevich P (1972) Shear modulus and damping in soils: measurement and parameter effects. J Soil Mech Found ASCE 98:SM6 603–624
Hayes G, Harley EP, Benz M, Wald D, Briggs RW (2011) The USGS/NEIC earthquake response team. 88 hours: the U.S. Geological Survey National Earthquake Information Center response to the 11 March 2011 Mw 9.0 Tohoku earthquake, Seismological Research Letters, vol 82(4). July/Aug 2011, pp 481–493. doi:10.1785/gssrl.82.4.481
Jeffreys H (1970) The Earth, 5th edn. Cambridge University Press, Cambridge
Malischewsky PG, Scherbaum F, Lomnitz C, Tuan T, Wuttke F, Shamir G (2008) The domain of existence of prograde Rayleigh-wave particle motion for simple models. Wave Motion 45(4). doi:10.1016/j.wavemoti.2007.11.004

Chapter 2
The Great 1960 Chile Megaquake

Abstract Descriptions of witnesses who experienced the great Chile megaquake may support the idea that prograde ground motion contributes to earthquake damage. Coastal changes in megaquakes may be related to a distinctive tectonic framework.

2.1 A Great Earthquake

On a bright Saturday morning in May, 1960, Chile awakened to a devastating earthquake of magnitude 7.5 that killed more than a hundred people in the city of Concepción, 400 km south of Santiago. Train service and commercial flights to the disaster area were cancelled. It turned out that this earthquake was merely a foreshock to the largest megaquake ever recorded: a main shock of magnitude 9.5, with epicenter about 300 km to the south of Concepción, known as the great 1960 Chile earthquake. The epicenter was near Mocha Island, off the coast of southern Chile.

Sunday, May 22, 1960, 3:15 p.m. The next day after the Concepción earthquake, some seismologists hitched a ride from Santiago airport on a military aircraft. They landed at Concepción airfield in time to experience the largest of all earthquakes. They were walking along the fence of the airfield when they noticed some parked aircraft rolling slowly back and forth. Then the cars parked outside the airport began to rock and some trees down the avenue tilted. A major earthquake was in progress.

Everything seemed animated by a silent slow motion. The earth was swinging back and forth in a leisurely rhythm of 2 or 3 s period. No earthquake of this size had ever been observed anywhere and no bigger one has occurred since.

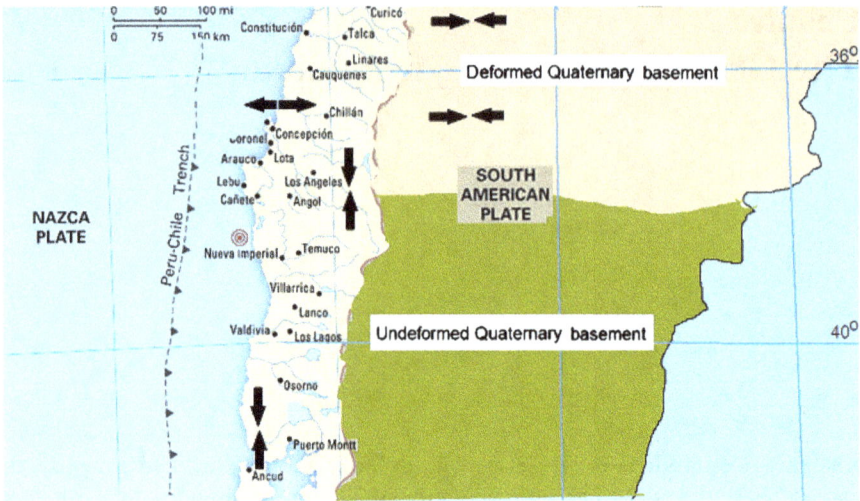

Fig. 2.1 Geological scenario of the 1960 Chile earthquake (bull's eye symbol), after Lavenu et al. (2002). The *arrows* indicate regions of predominantly compressive or extensional deformation

The magnitude was 9.5 on the Richter scale and the offshore Nazca-South America Plate boundary was torn over a 1,000-km fault rupture (Fig. 2.1). Estimated casualties numbered 5,700 dead, many of them reported missing in the tsunami.

There was no possibility of proceeding any further south from Concepción. The Nazca-South American Plate boundary had ruptured and generated a huge tsunami wave which rose to the unprecedented elevation of 30 m at some coastal locations.

In the Bay of Valdivia, wave heights of around 20 m were reported. The water depth in the open ocean is about 4,000 m; water waves propagating on this depth of water can attain about 800 km/h, roughly the speed of an airliner. Thus it is easy to compute how long it will take a tsunami wave to cross the Pacific Ocean from an epicenter in Chile. It can take the better part of a day. Unfortunately the 1960 tsunami caused victims and severe damage in Hilo, Hawaii and in Japan, because of insufficient warning on the arrival times of the tsunami.

In the epicentral region, a major landslide blocked the San Pedro River upstream of the city of Valdivia. Engineers from the Chilean Power Corporation (Endesa) recognized the impending danger and led a concerted effort to excavate a channel across the landslide allowing the water which was backing up at the landslide to be released gradually. But the earth moving operation proceeded slowly and virtually stopped when the equipment bogged down at the onset of the rainy season. The landslide was overtopped on June 23, 1960, less than a month after the earthquake. The flood wave traveled downstream and caused some additional damage to Valdivia but fortunately, no new victims were reported because of timely warning to the population.

The number of casualties in the 1960 megaquake was relatively low as the area was sparsely settled at the time. Many homes were built of wood and resisted the earthquake. Many victims were caused by the tsunami. After the earthquake, a building code was introduced for the whole country, and soft ground was recognized as a major factor of damage. For reinforced-concrete frame construction the code recommended the adoption of shear walls—a new technology at the time. A shear wall is a thin vertical structural element made entirely of reinforced concrete. Experience has shown that the introduction of shear walls in Chile paid off in terms of added seismic safety.

The main lesson of this earthquake was the large number of failures in foundations and embankments. A direct effect of the 1960 earthquake was the creation of the International Tsunami Warning Center (ITWC) in Honolulu, with the participation of most nations around the rim of the Pacific Ocean including all Latin American nations which share a coast on this ocean. The warning center is administered by the US National Oceanic and Atmospheric Administration and the Latin American countries are usually represented by their respective Navies.

2.2 Earthquakes and Coastal Geomorphology

Close to the epicenter of the giant 1960 Chile earthquake was the fishing town of Puerto Saavedra (about 14,000 inhabitants). This village on the estuary of the Imperial River was founded in 1885 as a frontier outpost after the Indian Wars. A campaign conducted by General Cornelio Saavedra had ended with the total defeat of the Mapuche Nation. Because of its exposed location at the mouth of the Imperial River, Puerto Saavedra was totally razed by the tsunami less than an hour after the 1960 earthquake. Most villagers were able to reach high ground but about 50 people drowned or were missing. The second or third wave was the largest. Successive waves continued hitting the town until late in the evening.

Puerto Saavedra is an outlet for agricultural and forestry production in an impoverished area of mostly indigenous population. Prior to 1960 the lower course of the Imperial River was navigable but the river shortened its course during the earthquake so that the estuary shifted its position from point A to point B as a result of coastal subsidence (Fig. 2.2). The fishermen at Puerto Saavedra survive by catering to summer guests and vacationers. Formerly they had access to the ocean but now the strong currents and tall breakers of the new estuary are too dangerous. Also, the port is no longer fronting on the river but on a lateral lagoon.

Lake Budi, south of the mouth of the Imperial River, is a brackish allogenic lagoon. It connects with the ocean through a narrow tidal channel known as the Budi River, with a length of about 15 km. The 1960 tsunami invaded Lake Budi and the water level in the lagoon rose about 2 m above the pre-1960 level. This represents the amount of coastal subsidence. The lagoon itself was recognized as a

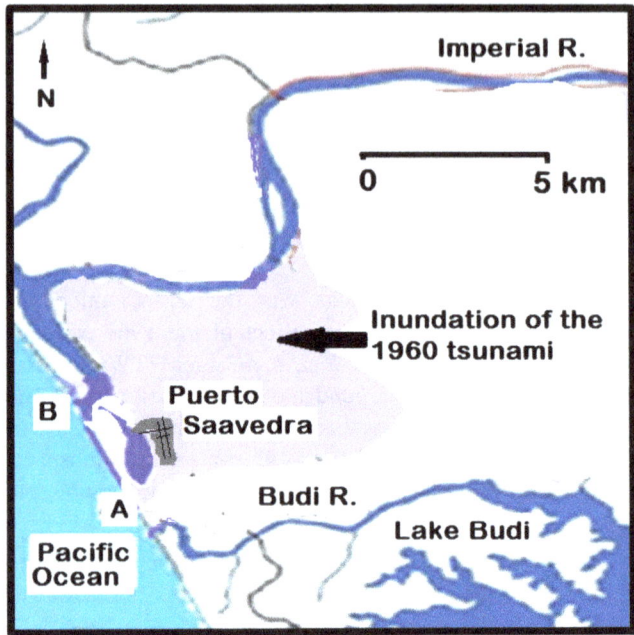

Fig. 2.2 Allogenic lagoon formation in the estuary of the Imperial River, southern Chile. *Shaded*, area of 1960 tsunami inundation; *A*, location of pre-earthquake estuary; *B*, present estuary. The Budi River, now a tidal channel, was once the lower course of the Imperial River

meander of the Imperial River that was cut off from it by some prehistoric earthquake (Lomnitz 1970, Wallner 2007).

The coastal configuration of southern Chile cannot be understood without investigating the geologic history of the area. The back-arc tectonics changes at the 38th Parallel (Fig. 2.1). To the north we find the strongly folded Sierras Pampeanas; active volcanism is absent. But south of 38° latitude the basement shows little deformation, active volcanism reappears and transpressional tectonics dominate in the magmatic arc. Wallner's drilling campaign in Lake Budi produced sedimentary cores that indicated a sudden change of regime from freshwater to brackish, about 2000 years before present. This finding confirmed that the lake had formerly been part of the lower course of the Imperial River: it was cut off by the receding coastline. Lake Budi was created by a large seismic event dated at about 2000 years ago. This "Year-Zero Earthquake" was a megaquake, of similar or larger magnitude than the 1960 earthquake.

No other information is available about this event. But it suggests that a systematic coastal subsidence of at least 2 m has occurred repeatedly on this stretch of coast at a possible time interval on the order of twenty centuries. The process of lagoon formation in the Year-Zero Earthquake was repeated in 1960 at Puerto Saavedra, where the river shortened its course once again and created a brackish lagoon. Another example of coastal recession in megaquakes is the 2011 Tôhoko earthquake, when the coast of NE Japan subsided by up to 1.2 m.

2.3 An Observation of Long-Period Surface Waves on Soft Ground

In 1968, 8 years after the great Chile megaquake, a visit to the epicentral region provided a new observation of the amount of coastal subsidence. Using some pre-1960 aerial photographs and taking the water level in Lake Budi as a proxy of mean sea level it was found that a permanent subsidence on the order of two meters had occurred. Boroa Mission, inland from Lake Budi and a two-hour drive from the provincial capital of Temuco, was used as a base of operations.

Boroa Mission was founded by Capuchin monks in 1883, after the Indian Wars. An earlier mission at Puerto Saavedra had been overrun and burned by the Mapuche warriors. The late Rev. Juan Bautista Wevering, missionary priest at Boroa, was interviewed about his personal observations of the 1960 earthquake. Father Wevering, then in his Forties, was born near the Dutch-German border. He joined the Capuchin Order after the Second World War.

He claimed to have actually *seen* the seismic waves. His testimony was intriguing, especially because he was not aware that elastic waves are invisible to the human eye. This may have enhanced his credibility: he was not prejudiced by any prior information he might have received. On the day of the earthquake, Father Wevering said, he took his lonely Sunday walk to a favorite spot at the top of a NNW trending fault scarp which overlooks the coastal plain. At a quarter past 3 p.m., in clear sunny weather, he strongly felt the earthquake. But he was not afraid. He intently watched a collective phenomenon he described as "seismic waves" rolling inland from the general direction of the epicenter (north-north-west). "The waves kept coming, row upon row, toward where I was standing," he explained. "The earthquake was still in progress: it lasted a long time. Clumps of trees standing on some low hills in the plain would bend over as the waves passed by. It was a beautiful sight. The waves looked and behaved like water waves except that they swept over solid ground."

If we may accept the observation of Rev. J.B. Wevering as a sighting of hybrid elastic-gravity surface waves by a nonscientist, there are some peculiar features of his description which deserve comment. He watched surface waves propagating on soft ground over the sedimentary coastal plain, coming from the direction of the epicenter. The phenomenon struck him as orderly, not chaotic. An essential point in his description was the detail that the trees in the coastal plain tilted in the direction of propagation, "like water waves". These hybrid waves, which we may call *transelastic waves* for short, were intermediate phenomena between elastic waves and gravity waves. As the ground tilted into the direction of propagation the ground motion must have been prograde, as in water waves.

Prograde ground motion helps explain why the overturning moment generated by these waves can be so significant. Much of the damage to tall, rigid reinforced-concrete structures in the period range of 1.5–2.5 s must be attributed to trans-elastic waves.

References

Lavenu A, Cembrano J, Costa C (2002) Margin-parallel vs margin-orthogonal compression along a convergent margin: the Andes of Central-Southern Chile and Western Argentina. Proc Int Symp Andean Geodyn Toulouse 1:367–370

Lomnitz C (1970) Major earthquakes of Chile: a historical survey, 1535–1960. Seismological Research letters, vol 75(3), May/June 2004, pp 368–378. doi:10.1785/gssrl.75.3.368

Wallner J (2007) Jungkänozoische Landschaftsgeschichte am Lago Budi, Ph.D. thesis, unpublished, University Jena, Germany

Chapter 3
The 1967 Caracas Earthquake

Abstract The 1967 Caracas earthquake had important features relating to the practice of earthquake engineering. Legal decisions involving the accountability of engineers are reviewed. Failure modes in Caracas included pancaking. The story of how a man escaped from the 9th floor of a collapsing building and saved himself and all members of his family is presented.

3.1 Caracas in 1967

The Caracas earthquake of July 29, 1967, M 6.5, was a rare disaster that caused serious loss of life in a relatively less active segment of the Caribbean-South American plate boundary. The epicenter was located on the coast of Venezuela, less than 30 km west of the city. This segment of the plate boundary features right-lateral strike slip and does not normally produce tsunamis, though some tsunamis do originate from more distant epicenters on the subduction front of the Caribbean Plate.

Despite its relatively low magnitude about 240 people were killed in the 1967 earthquake. A similar but larger earthquake killed several thousand people on March 26, 1812, during the War of Independence against Spain. Caracas had no modern seismic ordinance at the time the buildings were designed. Neither had any of the Caribbean nations. Engineers in the Commonwealth nations used the British or US standards and analogous conditions existed in the French and Dutch Antilles. After the earthquake there was a sustained effort of supporting seismology and earthquake engineering in Venezuela.

In the Palos Grandes section of Caracas, a fashionable residential area built on soft, marshy ground—it was the site of a former lagoon—four 12-story reinforced-concrete apartment buildings collapsed with important loss of life. A man who

Fig. 3.1 Pancaking damage at Ahmedabad from the 2001 Gujarat, India earthquake. Modified after Murty et al. (2010).

lived on the 9th floor of one of the buildings was having an evening snack in the kitchen when around 8 p.m., the earthquake struck. The man grabbed his wife and his mother by the arm and rushed them out of the apartment. As they reached the staircase landing the lights went out. They ran down the stairs in total darkness while the remains of their apartment rushed past them in a torrent of water and dust. When they reached the street the earth was still shaking. A maid who was taking down some clothes from a clothesline on the roof escaped unhurt but most other tenants were dead.

Pancaking (Fig. 3.1) is a characteristic type of failure in reinforced-concrete buildings. It can be deadly. The heavy slabs collapse on top of each other; rescue is difficult because the slabs cannot easily be lifted without endangering the lives of people who may be buried underneath. The 9th floor resident owed his survival to a coincidence. The building had been erected in two stages. His apartment was in the new 12-story structure which shared the staircase with the older structure. It looked like a single structure but actually they were two adjacent buildings. The one which contained the 9th floor apartment was less rigid—possibly because the staircase provided some added rigidity to the other building. At any rate, the two structures moved out of sync as they had different frequencies of vibration. There was practically no clearance between the two structures. Thus the earthquake caused the two buildings to pound into each other and the newer structure lost the contest and collapsed. The man saved his life and that of the two ladies by unwittingly crossing into the older building when he stepped onto the staircase landing. He never realized that the staircase belonged to the adjacent building.

A Failure Mode & Effect Analysis (FMEA) can help prevent future earthquake damage at the time a structure or system is designed. Failure modes may be detected in an otherwise well-designed structure: this is not uncommon. When suitably identified and remedied, the detection of failure modes can be effective in preventing new disasters from occurring.

3.1 Caracas in 1967

The specific failure mode shown in Fig. 3.1 was eventually recognized and taken care of by engineers. In the 1967 Caracas earthquake and in the 1985 Mexico disaster, spectacular examples of "pancaking" were detected and engineers learned how to design buildings where this failure mode would not occur. Progressive or sequential collapse of one floor slab on top of the other is caused by failure of the vertical columns. The basic engineering problem is how to distribute the loads: this depends on the relative role assigned in reinforced-concrete frame structures to floors, beams and columns. When correctly designed, floor slabs are not supposed to carry the weight of the building, much less to resist an earthquake. This is the job of the skeleton of beams and columns, called the "frame". A frame structure works much like the skeleton of a human being: it carries the weight of the body. The joints between beams and columns must be tough and flexible at the same time: they must be able to absorb earthquake stresses. Floor slabs are merely supposed to distribute the dead load to the frame. But unless the slabs are firmly anchored to the columns they may break loose in an earthquake and pile up as seen in the figure.

3.2 Accountability and the Law

Thus failure mode analysis must take the history, the location, and the use of a structure into account. The apartment buildings that collapsed in Caracas could have provided safe dwelling in Buenos Aires or in São Paulo, where earthquake provisions are not an issue. They might have been retrofitted by inserting fluid-viscous dampers at structural nodes. There are many engineering solutions for strengthening the seismic resilience of existing structures. At present, Caracas has a strict building code with adequate earthquake provisions; what is more important, there are excellent architects and engineers and the level of observance of the building code is highly regarded.

But in 1976 the Caracas buildings were death traps. What does the law have to say about liability of the builders, the designers, or the real-estate agents? Dennis Mileti, a critic of current paradigms of disaster mitigation in the United States, comments:

> Pitting such hazard reduction adjustments against other, often conflicting, powerful societal forces ... [constrains] a more effective natural hazard adjustment [by allowing for] the decentralized character of the American system with its long list of involved actors, the low salience of natural hazards on most people's agenda until a disaster happens, the limited resources available for mitigation, and legal and economic constraints regarding restrictions on land use...
>
> The prevailing paradigm foretells increasing frustration because we have the knowledge to guide effective societal adjustment to natural hazards and we are informed about the societal constraints that impede our action, but dollar losses continue to increase. A shift in perspective is needed (Mileti 1996).

The concern is that "technological fixes" could backfire. But what are the legal constraints? Up to the twentieth century American judges consistently ruled that a defendant was liable for injury only when there existed a contract between the plaintiff and the defendant—the sole exception being when human life was "in immediate danger". And the law in other countries followed suit.

However, in 1916 Judge Benjamin Cardozo decided that liability does exist if a manufacturer is aware that his product "will be used by persons other than the purchaser, and used without new tests. There must be knowledge of a danger, not merely possible, but probable" (*MacPherson v. Buick Motor Co.*).

In *Paxton v. County of Alameda* the judge ruled that acceptable engineering standards are defined as "*that level or quality of service ordinarily provided by other normally competent practitioners of good standing in that field, contemporaneously providing similar services in the same locality and under the same circumstances.*" Thus the acceptability of an engineering design depends on local standards.

In *City of Mounds View v. Walijarvi* the judge ruled for the defendant, an engineer. He defined engineering judgment as "gained from experience and learning." It made a difference in terms of engineering responsibility whether earthquakes are uncommon or frequent in the locality. Allowances must be made for "situations where a certain amount of unknown or uncontrollable factors are common." In other words, disasters are legally expected to be unexpected.

In *Gagne v. Bertran*, the judge sought to narrow down the area of engineering responsibility still further. He ruled that when we hire an engineer we "purchase service, not insurance." The advice provided by an engineer does not entitle the client to claim damages when the result does not meet his expectations. There was no mention of compliance with building codes.

Mileti insisted that "we have the knowledge" and "we are informed" about natural hazards and what to do about them. But the courts are more intent on protecting experts from the fallout of their own mistakes and reminding us that there are knowledge gaps, and that there is a whole lot to be learned.

References

Mileti D (1996) On the current U. S. natural hazards paradigm. In: Hassol S, Katzenberger J (eds) Natural hazards and global change. Aspen, USA

Murty CVR, Dayal U, Arlekar JN, Sailendar K, Chaubey SK, Jain SK (2010) Seismically Deficient Structures. Engineering Lessons. Geospatial World http://www.geospatialworld.net/index.php?option=com_content&view=article&id=20376&Itemidf=1373

Chapter 4
The 1970 Peru Earthquake

Abstract The Peru disaster is an example of how different natural and social systems can couple to produce a major failure. Climate change may have intervened by destabilizing the high-altitude glacier that caused the zero-survivor obliteration of Yungay in 1970. Possible measures of prevention against siliar accidents are discussed.

4.1 The Santa Valley

On May 31, 1970 at 3:24 p.m. a destructive earthquake of magnitude 7.9 struck an area of northern Peru. The epicenter was located off the port of Chimbote but the heaviest damage occurred some 100 km inland, in the Santa River valley. It was the worst natural disaster in the history of Latin America: the number of casualties remains unknown but an often-quoted estimate of 100,000 dead seems possible. UNESCO fielded an emergency team of international experts to the stricken area.

The Santa River runs parallel to the coastline between two high mountain ranges known as *Cordillera Blanca* and *Cordillera Negra*. An active range fault runs along the base of the White Range, a granitic batholith. Mount Huascarán is the tallest mountain in the White Range and in Peru: it is crowned by a glacier at an altitude of 6,700 m, highest glacier in the tropics. High-altitude glaciers near the equator tend to be unstable. During the earthquake a mile-long block of ice was broken off the leading edge of the glacier. It tumbled down the mountain at high speed and buried all 22,000 inhabitants of the small resort town of Yungay (Fig. 4.1).

The earthquake did not occur on the range fault but on the boundary between the Nazca Plate and the South American Plate. The Nazca Plate is an oceanic plate that moves eastward against and under the South American Plate. This process is

Fig. 4.1 a Aerial view of Huascarán glacier showing the path of debris flow. **b** The avalanche banked, overflowed the moraine (*left*), branched out and covered Yungay (*foreground*). Most of the sludge flowed on toward the river. **c** The town of Yungay after the disaster. Servicio Aerofotográfico del Perú and U.S. Geological Survey

known as *subduction*. It causes more than 90% of all earthquakes. It also causes much of the earth's volcanic and tectonic activity. Thus, while the Cordillera Blanca Fault did not cause the 1970 earthquake there was indeed a connection between the fault and the earthquake: both were ultimately caused by the same subduction process.

The Santa River is fed by glaciers. It flows from south to north along the base of the snow-capped Cordillera Blanca. Figure 4.1a shows how the ledge of ice dropped down the steep slopes of Mt. Huascarán in nearly free fall for more than 1,000 m. The ice formed a torrent of sludge that traversed a high-altitude lagoon and picked up speed as it mixed with water, mud and debris. Past the lagoon the ravine heads straight for the Santa River: here the avalanche banked and overtopped a lateral moraine (Fig. 4.1b, foreground). The town of Yungay was in the path of the overflow. It was covered by up to 5 m of sludge.

The site of Yungay was on a sloping river terrace overlooking the Santa River. The seismologists stopped the car on the roadside and proceeded on foot. They were greeted by a desolate sight: everything was covered by a brown-yellowish muck as far as the eye could see. The carcass of a bus emerged from the wet mud: one could not walk on it. Four tall palm trees marked the location of the main town square of Yungay (Fig. 4.1c).

Standing at the cemetery, highest point of the buried town, and facing the impressive bulk of Mount Huascarán, there was no sign of life anywhere.

The cemetery occupied the site of a burial mound or *huaca* of pre-Inca age. A large statue of Christ seemed to reach out toward the town: it had barely been spared by the avalanche.

4.2 Geology of the Disaster

Yungay has been afflicted with disastrous detritus avalanches or *huaycos* (from the Quechua word *wayqu*) since prehistoric times. On 6 January 1725 an avalanche which may have been caused by an earthquake killed 1,500 people. On 12 January 1962 a debris flow from Mount Huascarán killed 3,000 people in Ranrahirca and narrowly missed Yungay. The 1970 disaster killed around 22,000 people in Yungay alone. There was no practical way to rescue the remains of the victims, and the government decided to preserve the site as a collective burial site and a disaster memorial.

The hazardous location of the town of Yungay in the foothills of Mt. Huascarán was always recognized. Avalanches are common in the glacial valleys of the Cordillera Blanca. The Cordillera Blanca granitic massif was intruded in late Miocene. It extends some 200 km NNW-SSE, parallel to the coast of Peru. It is the highest mountain range in Peru and possibly in the tropics worldwide. At 9° latitude Mt. Huascarán is closer to the Equator than the Himalayas. The normal range fault which runs along the foothills is moderately active: it has a left-lateral component. It is similar to the Wasatch Fault in Utah and is assumed to generate earthquakes of up to magnitude 7 but none of this size has been recorded in historical times.

Every year *huaycos* cause damage and casualties all over Peru. A slurry of ice, snow, rocks and liquefied soil races down the steep mountain canyons. The glaciers of Cordillera Blanca cover about 500 km^2 at present. They have receded considerably since the 1970 earthquake. The landscape is typical of glaciation: mountain lakes between deeply incised, U-shaped valleys bordered by moraines, with spectacular truncated spurs and valleys sliced off by faulting. Its challenge has attracted mountaineers of many nations. Patrolling the glacier should be able to detect the onset of instability and should help prevent disaster.

Extensive seismic damage also occurred in Huaraz, the regional capital founded by the Spanish in 1574 at an altitude of 3,000 m above sea level. Huaraz was among the oldest, most picturesque Colonial cities in Peru. It is located upstream from Yungay, on a river terrace overlooking the Santa River with a fine view of the Cordillera Blanca. Three-story adobe townhouses with heavy Spanish tiled roofs crumbled in the earthquake. The rubble blocked the narrow downtown alleys and streets. An estimated 10,000 people—most of the population of the town—died in the earthquake. At present the town is booming again and the population is estimated to exceed 140,000 people. Reconstruction of Huaraz failed to preserve the Colonial flavor of the town: it prides itself on modern construction and twentieth century earthquake codes. Unfortunately no one is preventing the urban area to spread across the path of future avalanches.

4.3 Climate Change and Communicative Action

Yungay ("warm valley" in Quechua) was settled in pre-Inca times, as the large *huaca* tumulus of the local Cemetery suggests. Human remains and traces of early agriculture have been dated to 10,000 years B.C. The burial mound was re-occupied by Inca settlers, as the site of Yungay commanded a strategic location. Its significance was not missed by General Andrés de Santa Cruz (1792–1865), Supreme Protector of the Peru-Bolivian Confederation. In his attempt to unify both nations Santa Cruz was opposed by the *Restoration Army* of Peruvian separatists supported by neighboring Chile. The Chilean armed forces played a dominant role in the battle of Yungay of 20 January 1839 which sealed the fate of the Confederation. Santa Cruz was utterly defeated and about 2,700 soldiers were killed on the battlefield. This victory also boosted the national consciousness of Chile.

Forty years later, the Chilean army defeated the Peruvian forces in the War of the Pacific. Lima, capital of Peru, fell to the Chileans and the military occupation extended into the back country. After defeating the Bolivians as well, Chile annexed the province of Antofagasta and acquired the Peruvian part of the mineral-rich Atacama Desert in the Treaty of Ancón (1883).

The prolonged military occupation of Peru (1881–1883) caused outbreaks of armed resistance in the Andean highlands including Yungay. After the departure of the Chilean forces guerilla warfare turned into an uprising against the landlords. In 1885 a charismatic local leader, Pedro Pablo Atusparia, headed a peasant revolt and overran Yungay with several thousand men.

The Huaylas Corridor—as the Santa Valley came to be known— thus acquired a specific regional and cultural identity. It controls the main route of communication between northern and southern Peru. It is also known as the "Switzerland of South America" because of the imposing landscape, the mild climate and the abundance of water. The region is economically self-sufficient and the people are enterprising and independent-minded. The risk of natural disasters has played a major role in the history of the region.

Glaciers are in retreat worldwide. Since the 1970 earthquake the size of glaciers in the Cordillera Blanca has shrunk by more than 15%. Mount Kilimanjaro, in Tanzania, has glaciers that retreat even faster but they are closer to the Equator. Receding glaciers are not limited to low latitudes: some high-latitude glaciers are retreating even faster. It seemed logical to attribute the 1970 Yungay disaster to climate change. The high-altitude glaciers of the Cordillera Blanca were recognized as unstable well before global warming became a subject of international concern. Some glaciologists have suggested that these glaciers may have started retreating as early as around the mid-nineteenth century. Of course, glaciers hardly ever stand still: they are continually moving downhill like rivers of ice. A glacier is

4.3 Climate Change and Communicative Action

actually defined as a mass of ice which flows over a landscape. The Intergovernmental Panel on Climate Change (IPCC) has grown confident that global warming is being caused at least in part by human activity. At present there is little solid evidence that the glaciers in the Himalaya are retreating due to global warming, and IPCC has retracted from an earlier prediction that they might disappear by 2035.

The leading toe of Huascarán glacier creeps slowly downhill. It never stops. For reasons of topography it breaks off from time to time causing avalanches, with or without intervening earthquakes or climate change. The periodic recurrence of *huaycos* from Huascarán and other tall mountains of Peru is documented since the Inca period all over Peru. Huaycos were occurring before the discovery of America. Most localities in the Santa Valley of Peru have suffered environmental disasters of one kind or another. Yungay was rebuilt after the earthquake: it now has a population of around 20,000. Huaraz was rebuilt exactly on the same spot where it was founded: it is threatened by huaycos as well as by earthquakes. The urban area has been allowed to overflow into gullies where debris flows are known to have occurred in historic times. People moved back to Yungay, or as close as they dared to the site where Yungay once stood. Because of improved construction methods, earthquake risk has become more manageable but the exposure to avalanche risk has increased.

A basic rule for the prevention of avalanche risk is periodic inspection of the leading edge of the glacier. People living at the foot of high-altitude glaciers are familiar with the risk, and a periodic surveillance routine should be proposed, approved and established. The physical properties of ice change with depth, due to the weight of the overburden. The bottom of a glacier is plastic ice, which enables the entire mass to slide bodily downhill. On the other hand, the top 50 meters are of brittle ice which tends to form deep cracks called *crevasses*, much feared by mountaineers. They betray the presence of shear stresses in the ice. When crevasses start to form parallel to the leading edge of a glacier it is time to act. Protective measures include the installation of *snow nets*, commonly used in Switzerland to protect downhill areas against falling rocks. A snow net is made of strong steel cables to catch the falling mass on steep slopes. *Snow fences* are vertical stakes planted in the path of an avalanche to slow it down and deflect the impact away from inhabited areas. Many different kinds of obstacles can be used to slow down an avalanche: this strategy might have saved lives in Yungay. Experienced engineers know how to deflect the debris flow and prevent it from banking.

The most important preventive strategy has to do with risk communication. Modern sociology has developed a theory of society based on communicative action (Habermas 1981). This fundamental idea is relevant to the modern theory of risk (Lumann 1993) based on a distinction between *risk* and *hazard*. Risk is hazard plus communication. It is socialized hazard. In a famous example, Luhmann suggested that the invention of the umbrella converted the *hazard* of rainfall into a *risky* decision, namely whether or not to carry it when leaving home. If you don't take it along you risk getting wet and spoiling your expensive suit; and if you do, you risk forgetting it someplace.

Thus risk has become an essential ingredient of decision-making. In the case of Yungay, the risk of earthquakes and avalanches involves an important set of decisions by social actors, including the population and the authorities. Should the site of Yungay be abandoned? Should it be restricted to areas not located in the path of an avalanche? Should the central government intervene or should decisions be left to the local authorities? Should avalanche insurance be offered and if so, by whom? What should be done about avalanche risk in Huaraz and other endangered localities? Should avalanche information be provided to the endangered population and if so, by whom? Should Mount Huascarán be declared off bounds to visiting tourists? Should it be patrolled, to monitor the stability of the glaciers?

After 1970 there has been a tendency of governments to leave such decisions to the citizen. Insurance is widely available and communication between people at risk is increasingly being circulated by the press and by social networks while the role of government is confined to a regulatory function. But the hazard is still there, and arguably the risk to rapidly growing population centers such as Huaraz is increasing.

References

Habermas J (1981) Theorie des kommunikativen Handelns. Suhrkamp, Frankfurt
Lumann N (1993) Risk: a sociological theory. Aldine de Gruyter, New York

Chapter 5
The 1985 Mexico Earthquake

Abstract This great earthquake belongs to the category of unexpected technical failures, beginning with the foundation of Mexico City in a lake in 1325 and leading up to the 2011 Japan megaquake. The "knowledge gaps" surmised by N.P. Suh had to do with engineering geology. Social effects of the disaster included a spontaneous spurt of solidarity followed by revulsion against denial by authorities. The evolution of legal decisions on accountability is discussed.

5.1 September 19, 1985, 7:19 a.m.

Disaster struck Mexico 4 days after Independence Day, 1985. The earth began to swing back and forth in slow tempo, like a ship in a light breeze. The period of oscillation was between 2 and 2 and a half seconds—and it went on and on. It was vastly more alarming than some large seismic events scientists had felt. The ground motion appeared to contain a single frequency, like a backyard swing.

The dominant period of two seconds in the ground acceleration is characteristic of soft-ground conditions. The same waveform also showed up in the great 2010 Chile earthquake (Fig. 5.1). The 1985 Mexico City disaster furnished proof that this type of ground motion can cause severe damage to reinforced-concrete structures with an elevation of more than six floors.

Mexico City mud is the dominant soil type in downtown Mexico City. It is a black, plastic, organic material often described as a volcanic lacustrine silty "clay"—a rather unusual term as its unit weight is 1.1, or barely 10% heavier than water, when common wet clays weigh 1.6–2.0 times the unit weight of water. The natural water content of Mexico City mud is 360–400%. Soil scientists would classify it as a *histosol*.

Fig. 5.1 Response acceleration spectrum of the 1985 earthquake at SCT in Mexico City (*shaded*), as compared with the Building Code before (1976 version) and after the earthquake (1987 version)

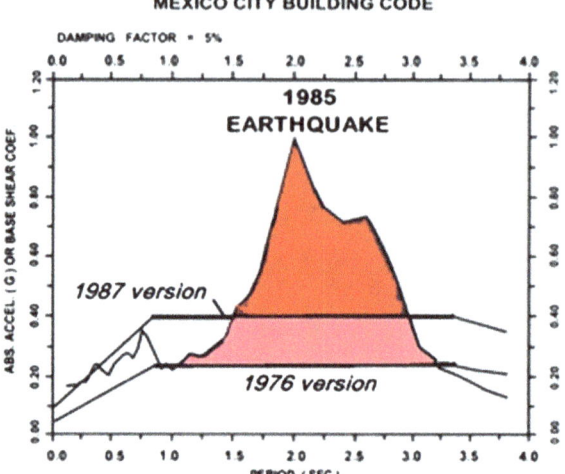

Diaz Rodriguez (1989) found that Mexico City mud behaves as an elastic material below 83% of the static strength. It features a closed hysteresis loop in this range. But above 84% of the strength the stress–strain response suddenly turns strongly nonlinear and there is rapid strength degradation in cyclic loading at a frequency of 0.5 Hz.

There was no generalized power blackout in Mexico City after the 1985 earthquake but TV and radio stations went off the air and the telephone went dead. Except for amateur radio the city was cut off from the rest of the world for the first twenty-four hours. At the university a preliminary location of the epicenter was performed using a locally available signal. The magnitude was estimated at 8.1 and the direction of the seismic waves at Mexico City pointed to an epicenter near the industrial port of Lázaro Cárdenas, on the Pacific coast of Michoacán state. This particular segment of the Cocos-North American Plate boundary had been inactive for many years and was suspected to be incapable of generating a damaging earthquake. The epicentral distance was about 400 km from Mexico City.

An aerial survey was arranged immediately, accepting the timely loan of a single-engine CESSNA piloted by its owner. Prompt information from the stricken area is a vital asset after a major earthquake, as large subduction earthquakes often cause more damage at locations inland than they do in the immediate epicentral region. This is partly due to geology. Mexico has a coastal batholith—an intrusive granitic body underlying the Pacific coast. Most coastal cities and beaches between Puerto Vallarta and Tehuantepec are built on this solid basement material that effectively mitigates earthquake damage. A similar situation exists along the coast of Chile.

Aerial surveying of Western Mexico between the Pacific coast and Mexico City revealed a puzzling feature of the 1985 earthquake. The shaking at Lázaro Cárdenas, in the immediate epicentral region, produced about the same peak ground accelerations as in downtown Mexico City, 400 km away, but the effect on structures was completely different. Something was wrong with peak ground acceleration as a measure of earthquake intensity. But it turned out to be a problem of physics, not of measurement theory. The damage in Mexico City was produced by surface waves with a period of 1.5–3 s, and these peculiar waves seemed absent in the epicentral region. More accurately, they lurked in the background. People had never really worried about slow, long-period waves and their possible effect on a type of modern structures.

This insouciance was reflected in building codes. Figures 5.1 and 5.2 (*bottom*) show that the 1.5–2.5 s period was not covered. It was unexpected in 1985 and remains unexpected today.

A small aircraft can be an invaluable research tool in such critical moments because of the breakdown in communications that appears to be a common feature of all disasters.

5.2 Lázaro Cárdenas

Few seismologists have actually experienced a large earthquake. This circumstance isolates seismology from other sciences. The opportunity of surveying a magnitude 8 event from the air within hours after it occurred should not be missed. Fortunately there are hundreds of flying clubs all over the world where private pilots will gladly volunteer to ferry a stranded scientist in an emergency. This happened after the 1985 Mexico earthquake.

Lázaro Cárdenas is an industrial city in Michoacán state; it is located at the estuary of the Balsas river. In 1985 it had less than 100,000 population. It was renamed after President Cárdenas in 1970. It is a major deep-water seaport. The town sustained significant damage in confined brick-and-concrete construction, a type of low-cost building type where gaps are left in a brick masonry wall for reinforced-concrete columns to be filled in later. The reinforcing bars used in this type of construction are often skimpy. The damage was assigned an intensity of VII on the Mercalli Scale.

The port area is sited on a flat sandy island in the estuary. There are important industrial installations including a steel foundry and a fertilizer plant: these were found to have sustained some structural damage but no major collapse. Rail tracks running along the beach were twisted into S shapes—a spectacular form of damage which corresponds to intensity VIII on the Mercalli scale. Two people were killed in a highway crash caused by loss of steering during the earthquake. On the whole, the effects of the earthquake in the epicentral area were less serious than one might have feared.

When the TV signal came back in the early evening of September 19 there were news pictures from downtown Mexico City showing total collapse of high-rise buildings and hospitals, cars buried under heaps of bricks, overturned TV antennas—a nightmare. Hundreds of volunteers were hopelessly scraping at the rubble trying to rescue victims of pancaking failure in reinforced-concrete-frame

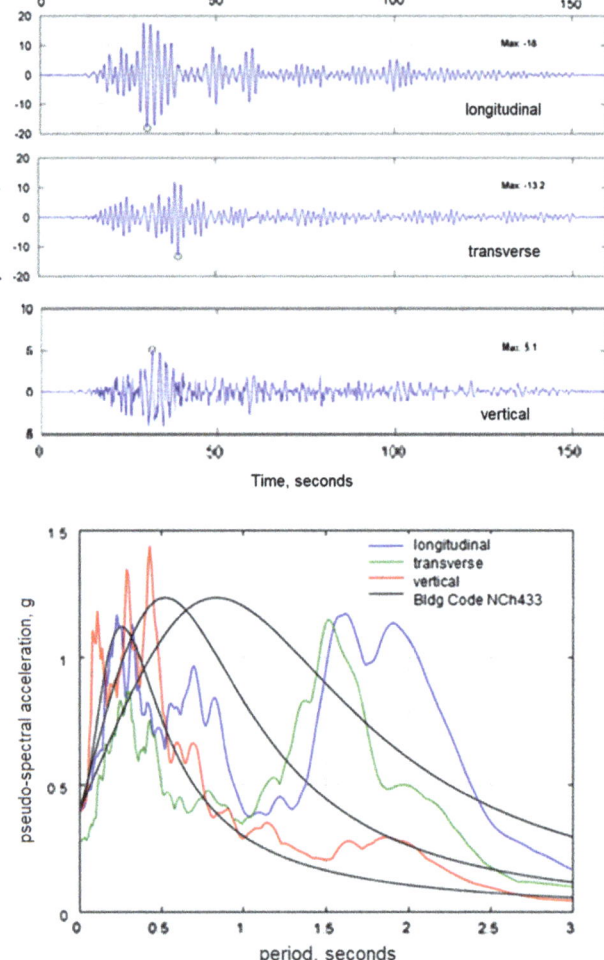

Fig. 5.2 2010 Chile earthquake. *Top*, filtered strong-motion displacement record (0.33–80 Hz) from downtown Concepción. Long duration and beating wave packets are as in Mexico City. *Bottom*, response acceleration spectrum for the same station (damping ratio = 5%) as compared to Building Code NCh433 for soil types II, III and IV. Notice the similarity with Mexico City (Fig. 5.1), Both earthquakes amply exceeded the respective building codes in the 1.5–2.5 s period range. Courtesy R. Boroschek

buildings. The airborne survey led to the conclusion that the disaster was mostly confined to downtown Mexico City as no other major damaged area was detected.

Downtown Mexico City is a densely populated urban area of about 100 square miles extent on the site of a former lagoon at an elevation of 2,240 m. It was known to experience high intensities from coastal earthquakes and the Building Code assigned higher peak ground accelerations to this area. For a magnitude 8.1

subduction earthquake at a distance of 400 km the pattern of damage was unusual. It suggested the presence of a waveguide between Mexico City and the subduction zone.

About 400 buildings of 7- to 18-floor elevation collapsed in the downtown area with very high loss of life. The damage was highly selective in terms of surface geology. It was felt much less severely in the volcanic hills that surround the former lake area: actually not a single building was reported to have collapsed on hard ground.

5.3 Mexico City

The Mexico City metropolitan area is a closed hydrological system. Water supply still depends importantly on ground water extraction and the geology of the low-lying districts is known mostly from the numerous water wells. In the Tertiary central Mexico was covered by the sea; the Cretaceous limestone is found approximately at sea level. The large Pleistocene volcanic edifices of Iztaccihuatl (inactive) and Popocatepetl (active) form a north–south barrier to the east of the city. Drainage to the south is blocked by the Ajusco-Chichinautzin monogenetic volcanic field which emerged in the Pleistocene. A major eruption occurred around 280 A.D. from Xitle, the most recent of more than 200 cinder cones, when a 15-km long massive lava flow buried the present southwestern area of Mexico City. The western hills are residential districts known as "Lomas"—they are underlain by massive welded tuffs locally known as *tepetate*.

In the low-lying flat districts which were formerly waterlogged damage to 7–18 floor structures was exceptionally severe. These structures had been designed under a building code that contained explicit earthquake provisions: the dominant reaction of engineers was one of unbelief.

The American continent was settled after 40,000 B.C. by migrant tribes that crossed the frozen Bering Strait from eastern Siberia. In the Twelfth century A.D., waves of nomadic Aztec tribes from the North infiltrated the high-altitude lakes of lush Anahuac Valley but were prevented from settling by native kingdoms that had made their home around the lakeshore. In 1,325 the Aztec legend claimed a favorable omen of an eagle perched on a prickly-pear cactus (*Opuntia ficus-indica*) devouring a snake on an unclaimed island in the lake. Tenochtitlan, the great capital city of the Aztec Empire, was founded here.

The city was connected to the mainland by four causeways with removable wooden bridges, making it impregnable to surprise attacks. In 1,521 a detachment of Spanish soldiers under Hernán Cortés besieged and razed the city. It was rebuilt in the Spanish style and renamed Mexico City. The Spanish drained the lake and the city extended beyond the original island into the soft, black, compressible lake bottom beyond. This soil type known as *histosol* is found in swampy or water-logged areas. The density of this soil is about 1.1 and sound propagates at a speed of 1,500 m/s as in water. Shear wave velocity is less than 50 m/s and Poisson's

Ratio is 0.499. Surface waves propagate on this material with a period of 2 s and a wavelength of 100 meters: in large earthquakes they can be seen with the naked eye.

The earliest report of sighting these waves may be found in the reminiscences of an eyewitness, García Cubas (1904). His description of the earthquake of June 19, 1858 (M 7.7) that caused severe damage in Mexico City is factually accurate and impressive. The author was riding in his coach along the old aqueduct—present-day Chapultepec Avenue—in the direction of Salto del Agua, when the earthquake surprised him near what is now the *Condesa* District. He alighted from the coach and observed the water in the irrigation ditch along the road shoaling and running in opposite directions rhythmically. The huge masonry aqueduct was winding on the ground like a huge snake as jets of foaming water spouted from it at intervals of about 100 m.

Soft-ground conditions were known to Colonial builders for many years. The Baroque architect Lorenzo Rodriguez (1704–1774), famed for his great Sagrario Temple on the Main Square, made allowances for subsidence. He provided an unusually tall portico in the façade of the church of La Santísima. The church was designed to subside by 4 m yet remain visually pleasing. Small windows in the bell tower helped provide an optical illusion. The church was undamaged in the 1985 earthquake but the City government decided to dig around the structure down to the original street level, thus defeating the intention of the architect. Hundreds of Colonial churches and palaces also subsided and survived the 1985 earthquake. They are enduring monuments to the foresight of their builders.

5.4 Social Effects of the Earthquake

An estimated 100,000 volunteers converged spontaneously on the downtown area to participate in the rescue operations. Housewives prepared meals for the volunteer rescue workers; medical students set up first-aid stations next to collapsed buildings; engineering students provided assistance in identifying buildings unfit for occupancy. For 3 days and three nights the rescue operations went on without interruption, and dozens of victims were rescued alive. A group of volunteers calling themselves Moles ("Topos") risked their lives by squeezing into collapsed buildings to reach people trapped inside. Sixteen newborn babies were rescued alive from the maternity ward of a collapsed hospital.

Valuable rescue work was also performed by non-governmental organizations, professional groups, foreign rescue teams, and the Mexican Army. Looting was not a major problem, as it was after the 2010 Chile earthquake.

The Tlatelolco project (102 apartment buildings, 12,000 apartments) was severely damaged in the 1985 earthquake. It was a showcase project of low-income public housing of the 1960s. Eleven buildings were demolished and about fifty were retrofitted. The Nuevo León Building, a 15-story apartment building with a length of 100 m, was a total loss. Two of its three sections capsized and at least 472 people were killed.

Tlatelolco currently serves a population of around 100,000 tenants. It is sited on 500 acres of former lake bottom soil near the northeastern end of Reforma Boulevard. The site was originally a railroad yard and a shantytown. Architect Mario Pani (1911–1993) and earthquake engineer Emilio Rosenblueth (1926–1994) designed and supervised the project. A floating foundation design first used at Tlatelolco used shallow boxes of reinforced-concrete to provide buoyancy.

Problems were apparently caused by apartment owners using the roof areas for building one-room dwellings which provided an added source of income. The foundation boxes of Nuevo León Building cracked and filled with water due to differential settlement. They lost buoyancy and the structure developed serious settlement damage. The tenants protested and eventually the building was retrofitted by providing a concrete mat at ground level extending outward from the original area of sustentation. This mat was supported by piles.

In the earthquake of September 19, 1985 the building was winding like a snake. Surface waves with a wavelength of around 30 m and a period of about 2 s caused the failure. Two sections of 33 m length capsized but the other third remained standing: it showed asymmetrical cross-bracing which may have contributed to the collapse.

5.5 The Senate Hearings

After the earthquake the Subcommittee on Science of the US Senate held a hearing to determine how the Mexico disaster might affect strategies of earthquake damage prevention in the United States. Professor Nam P. Suh of M.I.T. appeared before the Subcommittee on behalf of the National Science Foundation on October 3, 1985. He pointed out that influential Mexican engineers had been educated at US universities and that many of the structures which failed in the earthquake were essentially similar to some built in the US.

There were some obvious differences between the two nations but Professor Suh thought that they were not relevant as far as engineering design was concerned. On the contrary, the Mexican earthquake raised some pointed questions as to whether soft-ground conditions similar to those in Mexico City might also exist in San Francisco, Seattle and other major American cities. Professor Suh concluded by suggesting that there might be "a knowledge gap" about earthquake damage and urged this gap to be closed.

Why did Mexico City suffer such severe damage precisely in modern engineering structures, not in centuries-old Colonial masonry buildings? This question, Suh suggested, might be relevant to the safety of important cities all over the world.

In response, the National Science Foundation provided a fund of four million dollars to encourage emergency research on the Mexican earthquake. The participation of American colleagues stimulated joint scientific research on earthquake studies in Mexico, and the Mexican National Research Council (CONACYT) funded grants in earthquake research. A considerable amount of new knowledge was gained on the mechanism of Mexican earthquakes and on the behavior of structures on soft-ground.

But there remains a knowledge gap, as Professor Suh had feared. Leading engineers in Mexico and abroad conceded that the hazard estimates made before the 1985 earthquaker had been exceedingly optimistic, and that the building code had underrated the violence of earthquakes in Mexico City. The code was upgraded in 1987 and has been revised at regular time intervals.

5.6 Disaster, Denial, and the Law

The Mexican earthquake has been of considerable interest to social scientists. It was the worst disaster in Mexican history, and Mexico City was the largest city in the world at the time. Thus the 1985 earthquake had important social, political, psychological and economical consequences.

Denial, a well-known psychological mechanism, played a role in the disaster. After major earthquakes it often happens that officials tend to deny or minimize the facts. This was the case after the great 1906 San Francisco earthquake and it has also occurred in Latin America. Chambers of commerce, city administrations and tourist bureaus can be tempted to dismiss the possibility that a severe earthquake might ever occur in their town.

In Mexico there have been important advances in disaster prevention as compared with the level of preparedness that prevailed in 1985. The City administration developed an early-warning system that was installed in some schools and critical facilities. Earthquake drills were organized every year on the anniversary of the 1985 earthquake. Hospitals have been rebuilt and emergency services were greatly improved. More significantly, the new building code deliberately encouraged the widespread use of steel-frame construction.

But building codes are sets of recommendations addressed to architects, engineers and builders that are not enforceable as such. In the US and British legal systems they can only be enforced if they are enacted as law by an authority. In Spain the building codes are enacted by a Royal Decree; in Mexico there are normally enacted by the Governor of a state after approval by the State Congress. Enforcement is the job of building inspectors from the local Department of Works. In Mexico the responsible architect submits a project or set of drawings to the Department for approval but usually an inspection in situ is only performed in case of a problem or on demand. The building code is intended to protect public health, safety and general welfare with a minimum of meddling by the state. In Europe and the US there are model building codes that have been adopted by all member states or municipalities in order to avoid excessive complexity in practice. In Mexico the Building Code of Mexico City is widely used elsewhere.

But there is a lingering suspicion that the lessons of the 1985 Mexico earthquake have not yet been fully assimilated. Some lags remain to be addressed. One major lesson was that earthquake risk tends to be underestimated. This phenomenon is based on psychological research: future disasters are rarely anticipated simply because some of the basic risks are poorly known or not fully understood. In 1985 the dominant reaction of scientists, engineers and administrators was unbelief. No one had anticipated the destruction of modern multistory reinforced-concrete-frame buildings; on the contrary, they were vaunted for their "anti-seismic" design. Some authorities persisted, in the face of the evidence, in claiming that no major disaster had occurred.

The disaster was caused by an unexpected resonance phenomenon in high-rise buildings, related with the appearance of low-frequency surface waves of very long duration at dominant periods of 1.5–2.5 s. These waves appeared to be associated with local conditions in downtown Mexico City. In the 2010 Chile earthquake similar waves were associated with soft-ground conditions. Maximum observed ground accelerations were about 0.25 g near the epicenter and at 400 km distance. Modern reinforced-concrete structures with shallow foundations were especially vulnerable. Structures with deep foundations did not collapse. Colonial structures which had subsided 4 m into the ground were spared. The disaster was unexpected. Authorities went into denial. Improvements in the local building code were made but they were "reactive."

What is meant by a building code being "reactive"? International experience suggests that modifications in building codes are made after every disaster. If such changes are increasingly frequent there is a cause for concern—it means that the code leaves something to be desired. Of course no building code can be evaluated until it has been tested in an actual earthquake. The Mexico City Building Code had followed the lead of US codes in providing a substantial ductility allowance, to encourage safe design practices. Actually the ductility allowance was often used by engineers as a "fudge factor" to allow some leeway for engineering judgment.

References

Diaz Rodriguez J (1989) Effect of repeated loading on the strength of Mexico City clay. In: Cakmak AS, Herrera I (eds) Soil dynamics and liquefaction. Computational Mechanics Publications, Southampton
García Cubas A (1904) El Libro de mis Recuerdos. Arturo Gracia Cubas, Mexico
Reid H (1911) The elastic-rebound theory of earthquakes. U Calif Dept Geol Bull 6(19):413–444

Chapter 6
The 2010 Haiti Earthquake

Abstract What happens when everything goes wrong? Haiti is an example of a disaster with too many causes. Disaster culture in a nation cannot rely exclusively on importing technology from abroad. There must be a local basis for development grounded in education.

6.1 An Unmitigated Catastrophe

On January 12, 2010 at 4:53 p.m. an earthquake of magnitude 7.0 destroyed Port-au-Prince (Pòtoprens, in Krèyol), and damaged two neighboring cities, Jacmel and Léogâne. The population of the metropolitan area was 3.5 million, and the earthquake killed between 85,000 and 220,000 people. The affected population in the damaged area was about 3 million. The epicenter was located offshore, 16 miles west of Port-au-Prince. The Enriquillo Fault, on which the earthquake occurred, defines a microplate along the Caribbean-North America plate boundary. No tsunami occurred after this earthquake because the movement on the plate boundary was lateral, not thrusting—as on the San Andreas Fault in California.

An earlier destructive earthquake had occurred on the same fault on June 3, 1770. Since that date, no major earthquakes were felt in this general area. Social memory of natural disasters tends to grow dim in 240 years, and large earthquakes in the Caribbean region are sufficiently infrequent to make the region especially vulnerable, as several generations may not have experienced any damaging earthquake. Insufficient awareness of earthquake risk may partly explain the striking absence of local earthquake regulations, and the unusually high rate of destruction of structures in the 2010 earthquake.

The metropolitan area of Port-au-Prince, capital of Haiti, contains about half the population of the country. The downtown area is a low-lying area along the Bay of

Gonaives, with silty, saturated soft-ground conditions. The higher-lying Pétion-ville suburb was not as severely affected as was the downtown area, which was subject to liquefaction and almost totally destroyed.

After the earthquake the population seemed stunned or overwhelmed. The dead were left lying in the streets. No authorities were in evidence—there were no emergency services, no police, no medical services. Most hospitals were destroyed. The MINUSTAH Headquarters of the United Nations Mission, a large 7-story structure, was condemned as one-half of the building had pancaked in a cloud of dust.

The situation was comparable to the effects of a World War II bombing attack—worse than any natural disaster the world could remember. Practically no prevention measures had been considered before the earthquake. Reconstruction and recovery proceeded very slowly and depended largely on outside assistance. The contrast with the Chile earthquake which would occur 6 weeks later was striking.

6.2 Possible Causes of the Disaster

The geophysical causes of the Haiti disaster were tectonic. The epicenter was on the active strike-slip plate boundary between the Caribbean and North America Plates which slices across the island of Hispaniola in the east–west direction. Average relative motion on this plate boundary is 2 cm/year. Magnitude-7 earthquakes on the Caribbean-North America plate boundary are rare but not unknown. The 2010 earthquake was shallow—the focal depth was about 10 km—but this was to be expected in this region. Unlike subduction earthquakes, strike-slip earthquakes move the plates horizontally against each other, and shallow earthquakes tend to focus the seismic energy in the epicentral region.

None of these features is surprising. Then what is the cause of the enormous destruction wrought by this earthquake? Risk is a product of two main factors, hazard and vulnerability. Hazard is defined as the probability of occurrence of a damaging event: in the case of Haiti the earthquake hazard is moderate. But the vulnerability is extremely high: soft ground, no building code, no prevention, and one of the highest population densities anywhere. This earthquake was pinpointed at the Port-au-Prince metropolitan area, with more than 3 million people squeezed together at 50,000–100,000 people per square mile. Visible waves were observed during the earthquake in the downtown area.

Earthquakes don't kill: buildings do. The published findings of engineers who examined the damage after the earthquake are disturbing: none of the homes that collapsed in the Haiti earthquake had been built to withstand an earthquake of any size. For the sake of comparison, the intensity of the earthquake of 17 January 1994, magnitude 6.7, was among the highest ever recorded. It hit a densely populated area of the Los Angeles metropolitan area. There were 33 dead. But building codes existed in California since 1933 and the building industry was strictly regulated.

The 2010 Chile earthquake had a magnitude of 8.8: it was the fifth-worst earthquakes ever recorded. There were 562 certified dead. But: Chile has had a stringent building code for more than half a century. By 2010 practically all urban construction had been built under some kind of earthquake regulation. Engineers in Chile are strictly licensed, and earthquake engineers are well organized. They have their own professional society which holds well-attended yearly meetings to keep abreast of developments in building technology.

Regulation is not everything, and the number of people killed in the Chile earthquake was still excessive. It could have been further reduced. But the housing conditions in Port-au-Prince were substandard. The brutal destruction in Port-au-Prince bears the imprint of transelastic waves.

Let us look closer. The aboriginal Taino people of Hispaniola lived in dwellings made of branches and palm leaves. They were not a product of advanced technology but they were suited to the climate. It is not likely that we should go back to living in grass huts; but this is why modern nations have governments that take responsibility for the lives and well-being of their citizens.

6.3 Disaster Culture in an Earthquake Country

Haiti had hardly any significant experience in earthquake-resistant construction. The 2010 earthquake produced not a single seismogram or accelerogram in Haiti, as there were no stations. Fierro and Perry (2010) inspected and documented hundreds of homes that collapsed in the earthquake because of the utter absence of earthquake detailing. This was true even for multistory buildings in confined masonry construction. Many structures had flimsy reinforcement bars that were not anchored at structural nodes. The concrete was often of substandard quality. No building code of any kind was used in construction. Apparently no licensing was required for engineers or builders. Field work done by visiting engineering teams after the earthquake has produced little new information because the structures that failed in Port-au-Prince were so obviously inadequate that not much could be learned from them.

> Among the principal underlying causes of poverty that threaten Haiti's development and long-term stability are: social exclusion, poor governance, inadequate access to education and other social services, and limited economic opportunities along with environmental degradation (Beat Rohr, CARE International, interviewed 1 year after the earthquake).

Should a lack of disaster culture be added to the list? Disaster culture is sometimes misunderstood as political knowhow in dealing with international aid agencies. But disaster is probably the least explored, researched and understood major aspect of modern life. Hazard is generally underrated, and so is the difficulty of recovery and reconstruction. Building against earthquakes involves more than building codes. It is a sophisticated field of engineering which needs to be developed locally, with considerable experience and an active research and development establishment. There are no shortcuts.

Consider the access to education in Haiti. Public education is a crucial factor which may enable an impoverished population to break the vicious circle of disaster and poverty. Bahia, Brazil has great cultural and economic similarities with Haiti. Its population comes from the same region in western Africa. In recent years Bahia has successfully developed an industrial sector—automobiles, chemicals, aerospace, canneries, pharmaceuticals—on the strength of a vigorous public education program. It is growing faster than Brazil as a whole. Is Haiti too poor to catch up? Its GDP per capita is only around $1,200 and average earnings are about $2/day. Foreign aid represents 30–40% of the national budget and there is increasing social and economic inequality.

Haiti was the first independent nation in Latin America and the earliest free community of African origin. Prior to independence, it was the most valuable French possession in the Western hemisphere. It supplied most of the sugar and coffee consumed in the world and it buttressed the economy of France through slavery. Most of the slaves came from present-day Ghana, Togo, Benin and Western Nigeria. They spoke Yoruba, Fon, Ewe and other African languages. Their religion was Vodou or Vaudou, a syncretistic animistic cult similar to Candomblé in Bahia. European colonial regimes introduced a peculiar bio-taxonomy of racial status. As an example, the offspring of a European who married a white-skinned mulatto or "*albino*" was technically labeled a "jumpback" (*Negro torna atrás*). These racial subtleties were not exclusively found in the Spanish regime: in other Colonial systems the discrimination was similar. An eighteenth-century popular art form called "caste painting" reflected an ironic view of Colonial society hiding behind pseudo-scientific enlightenment.

Two years after the French Revolution the Haitian Revolution broke out. In 1825 France recognized Haiti and the United States followed in 1862, after the Civil War. Slavery was abolished in one stroke together with the caste laws in the newly independent republics of Latin America: yet prejudice persisted beyond legal distinctions. No new industries were developed to replace the slave economy based on sugarcane.

In 1915 President Woodrow Wilson decided to occupy Haiti. It remained a US protectorate until 1934. President Franklin D. Roosevelt retained control of the finances of the country until the end of World War II. After withdrawal of the US in 1947 there followed 10 years of instability. François Duvalier, better known as "Papa Doc", ruled as a dictator until his death in 1971. He was succeeded by his son Jean-Claude ("Baby Doc"), who was overthrown in 1986. President Jean-Bertrand Aristide was deposed in 2004 by a combined intervention of the United Nations: the *United Nations Stabilization Mission in Haiti* (MINUSTAH). 8,940 military personnel and 3,711 police from 40 nations including Brazil, Chile, France, Canada and the United States were sent to Haiti.

Higher education did not thrive in Haiti. At present the University of Haiti has around 15,000 students, most of them in the Law Faculty. Research facilities are very modest. There is no sustainable program of scientific research.

Reference

Fierro E, Perry C (2010) Preliminary reconnaissance report on the 12 January 2010 Haiti Earthquake. BFP Engineers Inc., Van Nuys, California.

Chapter 7
The 2010 Chile Earthquake

Abstract History is what happens in-between successive disasters. Prograde, long-period surface waves emerged first in the 1960 Chile earthquake and reappeared in 2010. The importance of economic resilience and regional development is discussed. Megaquakes may occur repeatedly in the same region but never exactly in the same place or the same way. Exponentially rising disaster costs are not sustainable.

7.1 Another Megaquake

The 2010 summer vacations were nearly over. Under a bright cloudless sky, thousands of summer guests and their families enjoyed their weekend on the beach. Chile has more than 6,000 km of coastline, the longest continuous stretch of coast of any nation fronting on the Pacific Ocean. The prolific seashore is dotted with coves and sandy beaches. As the weekend began, vacationers packed their bags in anticipation of a long return trip home, and went to sleep.

Saturday, February 27, 2010, 3:34 a.m. Suddenly the ground gave way. The whole country was seized in the throes of a megaquake of magnitude 8.8. The epicenter was located off the coast of Maule Province in south-central Chile. It generated a huge tsunami and destroyed the city of Concepción. At least 525 people were killed, most of them in the tsunami. Many victims were reported as "missing". The true figure of casualties may remain forever unknown.

The 175th anniversary of Darwin's earthquake, which caused terrible damage to Concepción and raised the offshore island of Santa Maria, was on February 20, 2010, exactly 1 week before the earthquake. Both events were remarkably similar. If megaquakes tend to repeat along the same stretch of subduction coast, the example of south-central Chile is worth closer examination.

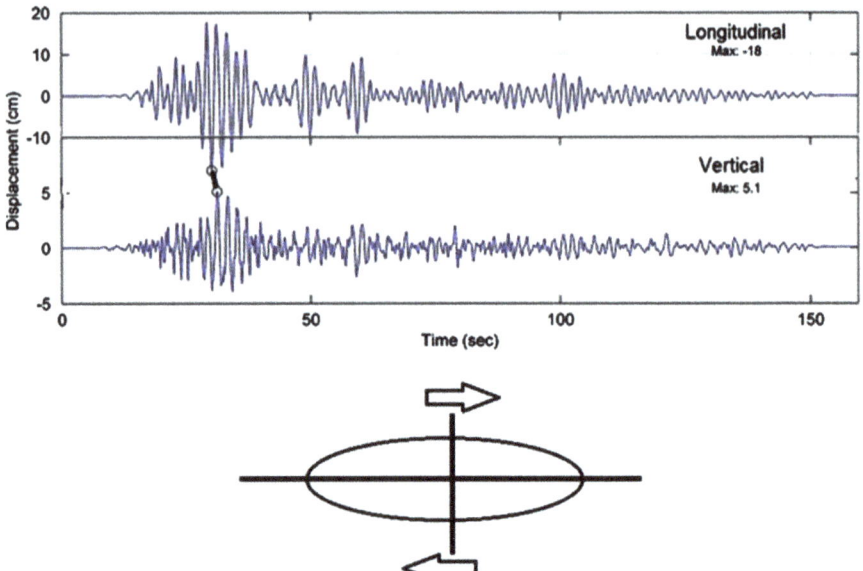

Fig. 7.1 Two-second period surface waves from the 2010 Chile earthquake recorded at downtown Concepción. The maximum longitudinal and vertical amplitudes are shown as small circles. Notice that the vertical component appears to be in advance of the longitudinal component, as would be the case in prograde motion. Courtesy R. Boroschek

The 2010 earthquake occurred past 3:30 a.m. on a Saturday. There was a nationwide power blackout between Taltal in the north and Chiloé in the south, including the capital Santiago and about 90% of users in the nation. Telephone and related communications were interrupted. The blackout was attributed to damage to the distribution network. Breakdown of communications affected the disaster response in critical ways (Cf. Chap. 1). Highway traffic to the disaster area was severely impaired and critical supplies (e.g., water and gasoline) could not reach the affected populations. Because of damage to police stations and related facilities there was no effective law enforcement in the major cities.

Outbreaks of violence included looting of supermarkets in Concepción and other coastal localities. The military had to be called in. Power supply was partially restored 2 days after the disaster but it took more than a week to reconnect the grid in the epicentral area. Two weeks after the earthquake 20% of users had no power and another nationwide blackout occurred. Much of industry was paralyzed.

Apparently because of severe damage to infrastructure the water supply network broke down in Concepción and other localities along the coast. High late summer temperatures and the perceived neglect of authorities caused tempers to flare. A curfew was enforced in the epicentral area.

Inevitably, comparisons with the 2010 Haiti earthquake were made. The *New York Times* concluded the next day that "while this earthquake was far stronger than the 7.0-magnitude one that ravaged Haiti 6 weeks ago, the damage and death toll in Chile are likely to be far less extensive, in part because of strict building codes put in place after devastating earthquakes."

7.2 Earthquake Damage Caused by Long-Period Surface Waves

Ground displacement in the plane of propagation was obtained in downtown Concepción from a filtered strong-motion recording. A two-second period surface wave is clearly seen on Fig. 7.1. Unfortunately it is hard to tell from such a record whether the motion was prograde or retrograde, as no rotational components are available. But the record does suggest the presence of prograde motion as the vertical component is slightly in advance of the longitudinal component.

The presence of prograde surface waves may mean that soft ground conditions in Concepción were such that the earthquake felt like waves on water. If the speed of shear waves was 50 m/s as in Mexico City the resulting tilt of the ground was about 0.2%. The result may be seen in the showy damage to twelve-to twenty story condominiums.

Experts have noted that only 3% of condominiums of 9+ stories were marked for demolition and only four collapsed in the earthquake. This refers to structures built after 1985. However most of the damage was attributable to a combination of soft ground plus large overturning moment in the earthquake.

Typical damage included many cases of buckled or crushed shear walls at or below grade. Many buildings lacked adequate foundations and/or subterranean levels. Local engineers tended to feel that the type and degree of failures was inadmissible. The PEER reconnaissance team report stressed that much of the damage was caused by relatively long-period ground motions that were not adequately covered in the building code (see Fig. 5.2). Also, condominium owners in Concepción and Santiago raised serious social issues that may have been underestimated.

7.3 Some Historical Sidelights

After the disastrous 1906 Valparaíso earthquake the Chilean government invited a distinguished French seismologist, Ferdinand Montessus de Ballore, to head a newly created Seismological Institute. Montessus de Ballore instrumented the early seismic stations and published some important research on earthquakes in Chile and Latin America.

Engineers in the 1940s developed an approach to earthquake design which was to be favored in Chile ever since. The principles of reinforced-concrete design were laid down by later engineers and scientists such as Rodrigo Flores, as follows:

1. Symmetry of design in order to control torsion in the horizontal plane,
2. Use of shear walls and other continuous elements with decreasing resistance upwards.

Later it was found that horizontal earthquake forces could not easily be controlled by the façade, the stair wells, elevator shafts and other rigid elements that carried small vertical loads. This insight led to the present introduction of outer and inner shear walls.

Early reinforced-concrete buildings were squat and rigid, with many degrees of freedom. Gradually more daring designs were introduced but basically the ideas were much more cautious than those that prevailed elsewhere in Latin America. This tendency was partly based on a peculiarity of the Chilean code. Up to some preset value of shear which depended on the quality of the concrete there was no requirement to verify that the rebars were able to take the diagonal stress. This meant adopting a relatively low shear stress—again, the solution was introducing more shear walls. The Japanese introduced ferrocrete as they realized the need for more ductility, and the Americans were also forced to lower their design forces with ductility. Finally, it was recognized that shear walls could be ductile too. But Chilean engineers were reluctant to abandon shear walls, and they paid scant attention to ductility. The Chilean model did not require the rigid nodes between beams and columns to be verified. This system is easier to build than frame systems that require much attention to detail if more ductility is to be attained.

Chilean engineers also noted that reinforced-concrete structures built in the Forties, Fifties and Sixties did exceptionally well in the 1985 earthquake. But the reasons for this good performance should be more closely examined. Before 1960 engineers designed with slide rules and ready-made computer programs had not yet been invented. They were not yet adept at cutting corners. People still used an earthquake design force of 10%–later increased to 12%–but we know that this is not high enough. Damage in roof structures has been the result.

The lessons of the 2010 disaster were visible even before 2010. The earlier model of disaster prevention cannot be sustained any longer because

1. The original Chilean system has been adulterated by introducing more shear walls.
2. Many new and possibly mediocre engineering programs have been introduced in the system of higher education.
3. Computerized design programs are ever more ubiquitous.
4. Designs of structures are adopted without compulsory revision.
5. Often new construction is not inspected.
6. The art of designing and using shear walls has been lost.

The 2010 earthquake has provided food for thought to engineers within and beyond Chile. There is a feeling that engineers may have been too conservative. New technologies are coming into widespread use and new factors of hazard are being recognized. The problem of long-period surface waves is an example.

7.4 Disaster and Regional Development

Acquiring resilience against disaster depends mainly on developing three objective aspects: (1) a robust economy, (2) a quality educational system, (3) a uniform regional level of development. Chile arguably had some initial advantages in all three aspects. However, in the 2010 earthquake there were far too many casualties, especially in the coastal region.

Chile is known for a high incidence of earthquakes, tsunamis and volcanic eruptions. Unlike Haiti, where over 220,000 people may have been killed in the 2010 earthquake, Chile did have frequent damaging earthquakes. The high incidence of earthquakes meant seismic stations, building codes, and an active program of earthquake research. However, the material losses attained similar levels as in Haiti. Dozens of small coves and beaches between Pichilemu and Puerto Saavedra were damaged or destroyed by a tsunami with waves as high as 11 m. The proportion of fatalities in coastal villages as against inland urban areas may have been as high as 3:5.

Chile's 4,000 km of coastline is an unevenly developed, partly impoverished region with serious problems of exposure to disaster. The traditional resource of this economic region was fishing. The marine resources of Chile's Pacific coast are amazing. They may be unique in the world. Valuable edible species include the giant mussel or *choro zapato*, the red conger eel or *congrio*, the *corvina* or Chilean drum fish (*Cilus gilberti*), the giant sea urchin or *erizo* (*Loxechinos albus*), the Chilean abalone or *loco*, the Antarctic king crab or *centolla*, the razor clam or *macha*, the piure (*pyura chilensis*), the cholga (*Aulacomya ater*), the giant squid (*calamar*), the scallop or *ostión*, the Patagonian seabass or *merluza negra*, the Southern kelp or *cochayuyo* (*Durvillaea antarctica*), and many species of crustaceans, cuttlefish and squid. Yet there prevails a surprising lag in marine biology and ocean studies, and social problems abound.

Only about 230,000 ethnic Mapuche remain in southern Chile: the others have migrated to major cities, especially Santiago. Poverty, disease, and discrimination prevail. The fishing industry was largely turned over to industrial interests that have shown little sensitivity for conservation. Overfishing and joblessness are endemic.

Chile is a member of the Organization for Economic Cooperation and Development (OECD) that—unlike Mexico, the other Latin American member of the OECD—does fairly well in the periodic reviews of public education and national science and technology policy. The Chileans' level of educational achievement is highly respected. However, the gross domestic product of Chile is only $14,811 per capita—below Mexico's $15,196; and in terms of income inequality, Chile ranks worst of all member countries.

Of particular relevance is a more energetic development of disaster science, an indicator of the degree of disaster preparedness of a country. In Chile this depends on the academic evolution of the University of Chile, plus a handful of other universities. Since 1958 the National Seismological Service has been entrusted to

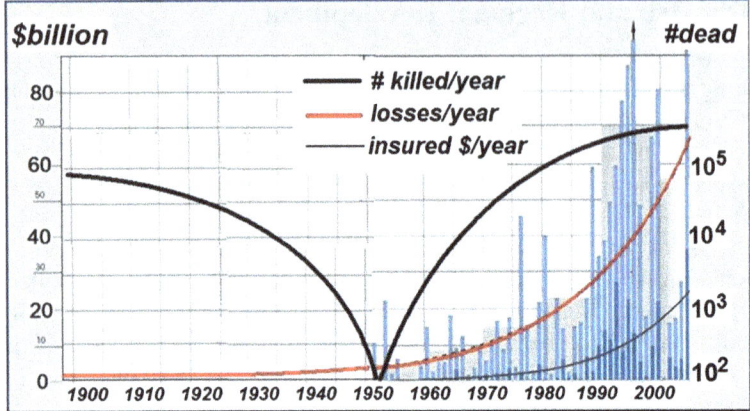

Fig. 7.2 World earthquake losses. The *light blue bars and red averages* represent total losses per year; the *dark blue bars and light black averages* are insured losses per year. The heavy black trends represent the average number of fatalities per year. Notice that earthquake losses tended to decay but rebounded after 1950

the Department of Geophysics at the University of Chile: basically, to the research department in seismology. The academic staff report on the country's numerous earthquakes to the National Emergency Office (ONEMI), the Navy Hydrographic Office (SHOA), and the press. Information is provided Mondays through Fridays within working hours, but the 2010 earthquake showed that this is not enough. Surely a national earthquake service should be available on a 24/7, round-the-clock basis.

A robust economy and a quality educational system are necessary but insufficient by themselves to guarantee safety against disasters. Chile had both. Rebuilding destroyed beachfront property in Talcahuano, Dichato, Iloca, Pichilemu, Constitución and other coastal localities is under way. The reliability of tsunami warning systems has been improved. But there is not enough independent, creative thinking about earthquakes.

There are two seemingly opposite trends in natural disasters, both attributable to technology: (1) a decrease in the number of casualties, (2) an escalation of disaster costs. In technologically advanced countries the number of casualties diminishes steadily. This has been credited to an implementation of effective public policies, assisted by appropriate scientific and technological developments. On the other hand, around 1950 the economic cost of disasters started to increase and continues to rise exponentially, year after year (Fig. 7.2). This exponential rise is not sustainable.

Quarantelli and Walter (1997) attributed the exponential rise in disaster costs to "ineffective policies, coupled with the failure to develop or adopt appropriate technologies." It seems, however, that explanations involving effective and ineffective policies at the same time cannot be correct.

Reference

Quarantelli E, Walter L (1997) Natural hazards and global change. In: Hassol SJ, Katzenmerg J (eds) The elements of change. Aspen, Aspen Global Change Institute, pp 1–80

Chapter 8
A List of Significant Earthquakes in Latin America

Abstract A review of significant earthquakes in Latin America is provided. The definition of a "significant earthquake" may vary from country to country. Latin America is a highly variegated region of high seismic risk.

8.1 Introduction

In 2010 the total population of Latin America was about 561 million. Latin American nations with a significant earthquake risk include the following 20 countries: Argentina (western), Bolivia, Chile, Colombia, Costa Rica, Cuba (eastern), Dominican Republic, Ecuador, El Salvador, French Guiana, Guadeloupe, Guatemala, Haiti, Martinique, Mexico (western and central), Nicaragua, Panama, Peru, Puerto Rico, and Venezuela (northern). The list includes 19 significant earthquakes in the period of 105 years between 1906 and 2010.

Notably absent in this list are Argentina (eastern), Brazil, Paraguay, Uruguay, and Venezuela (southern). These nations or regions do have some intraplate seismic activity: seismic risk is not absent anywhere in Latin America. But the deep interior of the South American Plate is not regarded as *significantly* seismic. Much of it is underlain by a large stable craton or tectonic shield known as *Amazonia*, of Precambrian age.

Table 8.1 Significant earthquakes in Latin America, 1906–2010

Location	Date	Magnitude	People killed	Damage (million US dollars)
Michoacan, Mexico	1911 Jun 7	8	n/d	45
Puerto Rico	1918 Oct 11	7.5	116	4
Managua, Nicaragua	1931 Mar 31	5.5	2,450	15
Oaxaca, Mexico	1931 Jan 15	7.9	n/d	n/d
Jalisco, Mexico	1932 Jun 18			
Chillán, Chile	1939 Jan 25	8.3	28,000	38
Quiches, Peru	1946 Nov 10		1,400	
Acapulco, Mexico	1957 Jul 28	7.5	160	25
Southern Chile	1960 May 22	9.5	3,000	880
Huaylas Corridor, Peru	1970 May 31	7.8	66,800	500
Managua Nicaragua	1972 Dec 23	6.0	5,000	800
Orizaba, Mexico	1973 Aug 28	7.1	539	
Motagua, Guatemala	1976 Feb 4	7.5	22,778	
Michoacan, Mexico	1985 Sep 19	8.1	10,000	4,000
Napo, Ecuador	1987 Mar 5	6.9	1,000	1,500
Off Nicaragua	1992 Sep 2	7.0	170	
San Salvador	2001 Jan 13	7.6	1,167	1,660
Port-au-Prince, Haiti	2010 Jan 12	7.0	210,000	300,000
Central Chile	2010 Feb 27	8.8	800	20,000

The above list may contain omissions as the definition of a "significant earthquake" depends on the criteria of significance, which may vary from nation to nation. The main conclusion that might be drawn from the list is that Latin America is a highly variegated region of high seismic risk (Table 8.1).

Chapter 9
Conclusions and Recommendations

Abstract This chapter contains thoughts on megaquakes as seen from the point of view of the engineer. Terzaghi's Six Commandments included a warning against cutting corners. Tsunamis could be caused by recoil of the subducted plate. Can disasters be "affordable" or "sustainable"? Some final considerations on sustainability, building codes, disaster culture, and ethics, are provided. A long list of recommendations is appended.

9.1 Summary

It is time to summarize. This small book was intended to provide an explanation of earthquakes—and earthquake disasters—in a modern Latin American context. We raised some holistic questions: What is the earthquake risk in Latin America today? Is Latin America facing a serious threat of earthquake proliferation? How can we escape the dilemma of sustainability?

Latin America is a huge geographical region and an important cultural entity. It is prone to natural, economic and political disasters: this is among its salient features. Earthquake hazard is far from evenly distributed over the region: Brazil, largest of Latin American nations, is not significantly threatened by earthquake hazard. But not even Brazil is entirely free of earthquakes.

Charles Richter used to say that "earthquakes always tend to recur in the same places, but never *exactly* in the same place." Richter meant to say that every earthquake is different from, but similar to other earthquakes. Earthquakes are unexpected, yet they are expected to be unexpected. Oscar Wilde professed to find this boring. Actually, earthquakes are not boring. They can be fascinating, except to victims.

9.2 Megaquakes

Megaquakes are very large earthquakes, so large they are beyond definition. Only six large earthquakes are generally identified as megaquakes: half of them—the 2004 Sumatra–Andaman earthquake, the 2010 Chile earthquake, and the 2011 Tôhoku, Japan earthquake—occurred early in this century. Two of these occurred in Asia and the other in Latin America.

There is no silver bullet against megaquakes. No superhero in a flowing cape will come to our assistance. Karl von Terzaghi (1883–1963, Fig. 9.1), a professor of engineering at Harvard, came close: he taught classes wearing a Superman cape.

Terzaghi taught his students to love engineering and hate disasters:

> Engineering is a subtle art: it enables you to reach correct results without bothering to climb the logic tree. It takes love of your subject—keen observation is as important as analysis. Poor assumptions produce terrible results.

Terzaghi was steeped in European and U.S. engineering tradition but his observations applied to Latin America as well. Poor assumptions have produced some terrible results. Thus the 2010 Haiti earthquake did not rate as a megaquake because it was "only" a magnitude 7.0 event—as if 220,000 dead were nothing. The press compared the Haiti with the Chile earthquakes because both were in Latin America and happened to be close in time: but this was the wrong conclusion. Nothing could be learned from the Haiti earthquake, an unmitigated disaster. On the other hand, the Chile earthquake evidenced more serious technical and societal weaknesses than did the Haiti earthquake.

9.3 Sustainability

The exponential growth of disaster losses is not sustainable because it contradicts the long-term maintenance of human wellbeing. It is true that wealth is also increasing exponentially but no human society can afford spending its wealth on increasingly costly disasters.

The world population has been rising exponentially for centuries and so has the size of the world economy. China, India and other emerging economies aspire to enjoy similar living standards as does the industrialized world. The question is: can we afford it? The holistic answer is *no*, because the level of consumption in the West is already unsustainable. Disasters are expressions of unsustainability.

China spends 2% of its GNP on research and development, less than do most developed nations. But Mexico is spending less than 0.4% of its GNP on science and technology, less than what we are spending on disaster assistance. Latin American nations seem to find disaster affordable, they spend much more on emergency assistance than they do on research and prevention.

Fig. 9.1 Karl von Terzaghi (around 1924). From Wikipedia commons

If we ask the insurance industry, disasters are affordable. More and more risk is transferred from governments to the private citizen. The idea is to allow the free market to regulate risk. The ideal society is a society where everybody is insured. But whatever the market does, human longevity does not increase exponentially. The maximum lifespan of humans is increasing very slowly, from about 100 years in 1900 to 122 years in 2011. The generational turnover is around 30 years. Middle-class homes and nuclear power plants are built to last around 40 years. Whatever our life expectation may be, all insurance ends up being life insurance.

9.4 Tsunamis

The average return period of megaquakes is on the order of centuries to millennia. There is a physical limit for the size of large earthquakes: it is the size of the planet. Significantly, as far as megaquakes are concerned, the highest loss of life is in tsunamis—the aspect of megaquakes we know least about.

A tsunami is a water wave—among the best-understood of physical phenomena. Yet the causation of tsunamis is not fully understood. Tsunamis are usually attributed to the mechanism of thrust earthquakes at plate boundaries called *subduction*. The origin of subduction is convective motion in the deep-seated mantle material that moves the subducting plate toward the subducted plate (Fig. 9.2a).

Both plates move to the right, away from the body of water. In Fig. 9.2b, the rupture causes the upper plate to recoil. *Recoil* is the backward momentum caused

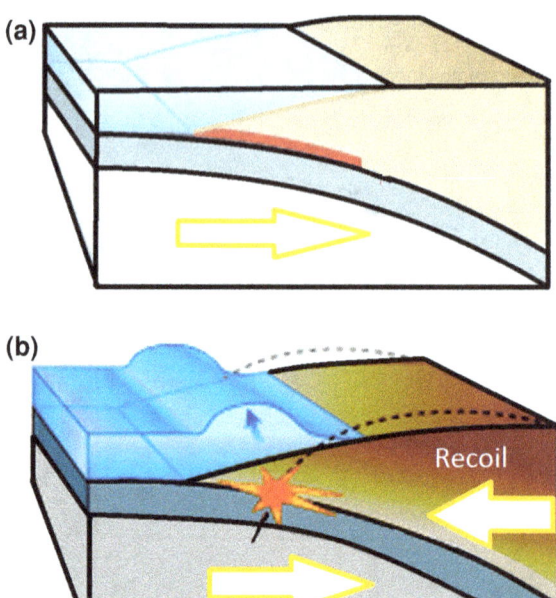

Fig. 9.2 A possible origin of tsunamis. **a** Mantle convection (*arrow*) causes a stationary stress to arise on the interface between the subducting and subducted plates. **b** The interface ruptures and moves (*lower arrow*) in a major earthquake. The free wedge of the subducted plate recoils (*top arrow*) and transmits kinetic energy to the water. Modified from graphics by Susan Mayfield and Sara Boore, courtesy of Geist EL, Gelfenbaum GR, Jaffe BE, Reid JA, Reducing the risk from coastal geologic hazards. U.S. Geological Survey, Fact Sheet 150-00

by reaction to a forward impulse according to Newton's Third Law, as when firing a gun. Momentum is transferred from the subducting plate to the overriding plate in the direction from left to right. Because of the wedged-shape overriding plate the power of the recoil increases from right to left. This power is transferred to the water wedge across the contact between the ocean and the overriding plate. The momentum in the water is exactly equal to the momentum of the earthquake.

This explanation accounts for the absence of tsunamis in strike-slip earthquakes. It also suggests that subduction earthquakes should always produce tsunamis. Small thrust earthquakes do produce small tsunamis but, as a rule, the momentum is not sufficiently large to generate a significant ocean wave. The tsunami signal is hidden in the noise of the surf. On the other hand, megaquakes cause large tsunamis.

9.5 Disaster Culture

In Latin America the term *disaster culture* has been widely used to mean different things. Local governments have used it to justify a policy of public earthquake drills once a year. If the earthquake does not occur the policy is successful; if it does, and there are casualties, they are blamed on a deficient earthquake culture.

The proof of the pudding is in the eating. Earthquake drills are useful: it helps to know that it takes us 4 min to reach the street from our 9th floor apartment. But by that time the megaquake will be almost over. The victim should have known better, and should have moved to a safer building. Instead of an earthquake drill, an inspection of housing conditions would have been more to the point. But the authorities are blameless: they survived.

Disaster culture means more than teaching children to crawl under a desk when an earthquake strikes. What if the home is made of flammable materials? It happens in Kobe, Japan in 1995.

Disaster culture can actually be important and effective in several ways. It can be imparted by a top-down approach. An earthquake drill can be useful to the participants if it sparks a reflection on urban systems, and how they react in disaster conditions. Downtown Kobe was a maze of narrow alleys: after an earthquake they would be filled with rubble, and the fire engines would not be able to get through. If disaster experience is to be incorporated in disaster culture, there are many ways of recycling this information to the citizens.

A valuable initiative will consist of installing an information bureau. Citizens should be able to find out about zoning, and which earthquake zone your home is in. Such information should be readily available to the public. It is comforting to know that engineers must have this information but ordinary citizens—men, women and children—should also have access to it. People need to know their exposure to hazard and how to make use of it.

Mexico and Chile are very different societies but both are in Latin America. Both are exposed to similar problems. A favorite housing system is reinforced-concrete frame construction of over six story elevation. But the social use of this system is slightly different. In Concepción, Chile, the collapsed buildings in the 2010 earthquake were condominiums. The tenants were middle-class people of modest means who could not afford to lose their property in an earthquake. The law made developers accountable for losses within a period of 10 years after construction. Yet not all of them would respond. In Mexico City the same buildings existed but the tenants were different. In the 1985 earthquake, the collapses were in low-income housing projects for government employees.

There is no substitute for disaster culture: it should be allowed to trickle down into the social fabric. It should permeate day-to-day decisions. But disaster culture should be based on accurate and reliable disaster information.

9.6 Building Codes

In Latin American universities the accreditation process for engineers is jealously guarded to preserve the system from intruders, especially from schools and universities from abroad. Academic inbreeding is not frowned upon. As an example, the National Autonomous University of Mexico (UNAM) retains the right to recognize or reject degree-holders from other universities or schools of engineering, whether from within Mexico or from abroad. This practice is not unique to Mexico: all countries have similar arrangements. In 1994 Mexico signed the

North American Free Trade Agreement and degree holders from many important foreign universities have since been recognized by the engineering profession. But Mexico has not yet signed the Washington Accord, an international agreement which recognizes the full accreditation and equivalency of undergraduate engineering degrees for all signatory members, including Canada and the U.S.

A building code is a guide to engineering practice written by engineers for engineers—with suitable restrictions. It is an instrument which may preserve, enshrine and sometimes modify the behavior of the engineering profession. Ideally, a building code is an engineering tool for the creative application of scientific principles to the design of structures able to resist earthquakes economically and safely. Such codes are sets of rules that specify the minimum acceptable level of safety for buildings and other structures. They may be adopted by cities, states, or countries. And they may be enforceable after they are enacted by the corresponding authority, e.g., by Congress or by the Governor of a State after passage by the state congress.

Each Latin American nation has its own legislation. Most Latin American legislations have a common origin in the Napoleonic Code, and ultimately in the legal system of Spain and Portugal, and in Roman Law. Latin American building codes are enforceable in theory but not necessarily in practice. The Chilean situation is an example. The country had years of experience with building codes: a nationwide code was enforced at the municipal level by building inspectors. But during the military dictatorship (1973–1990), accepted procedures for inspection by municipal authorities were modified. After democracy was restored, these modified procedures were preserved. Builders are entitled to appoint their own inspectors. They may be independent consultants or they may be self-employed or on the payroll of the builder, as was the case in the Alto Rio Building in Concepción which suffered a spectacular failure in the 2010 earthquake.

This is not to say that the present system is necessarily bad. It is not too different from the U.S. system. But it relies on the morality of the builders, all of them honorable men—no doubt of it. Nevertheless, there is a shadow zone where engineering meets business. This shadow zone is not regulated by law: it is dependent upon codes of ethical behavior.

9.7 Engineering Ethics

The engineering profession in most Latin American countries subscribes to basic principles of ethics. Engineering ethics is recognized as a separate field of research with its own journals. The study of engineering ethics pertains to the philosophy of science and technology.

Consider the code of ethics of the American Society of Civil Engineers in its 2006 version:

(1) Engineers shall hold paramount the safety, health and welfare of the public and shall strive to comply with the principles of sustainable development in the performance of their professional duties.
(2) Engineers shall perform services only in areas of their competence.
(3) Engineers shall issue public statements only in an objective and truthful manner.
(4) Engineers shall act in professional matters for each employer or client as faithful agents or trustees, and shall avoid conflicts of interest.
(5) Engineers shall build their professional reputation on the merit of their services and shall not compete unfairly with others.
(6) Engineers shall act in such a manner as to uphold and enhance the honor, integrity, and dignity of the engineering profession and shall act with zero-tolerance for bribery, fraud, and corruption.
(7) Engineers shall continue their professional development throughout their careers, and shall provide opportunities for the professional development of those engineers under their supervision.

This code of ethics is largely subscribed to by engineers in Latin America. Engineers are a special breed. Physicians have their oath; engineers don't need one. They are interested in results. But unlike physicians, they seldom deal directly with the end user. As a rule they work for another engineer who works for firms of various sizes. Various ethical dilemmas arise in the course of an engineer's professional activities.

If we examine the above code of ethics more closely we may discover that some points may conflict with others. As an example, an engineer may act as a *faithful agent or trustee of his employer* yet fail to hold paramount the *safety, health and welfare of the public*, and vice versa. Thus point (1) may conflict with point (4). Such ethical dilemmas may be relevant to the occurrence of earthquake disasters, as the engineer may be unaware of them. An earthquake can be an engineer's most honest and severe critic. These wicked natural events unerringly spot weaknesses in design or construction, and there is no arguing with an engineering failure.

Some prominent engineers have proposed their own versions of engineering ethics. Karl von Terzaghi (1883–1963) suggested the following "commandments of the Engineer" (Goodman, 1999):

9.7.1 Six Commandments of the Engineer

(1) Take up a challenge only if you know what you are doing.
(2) Assume the worst case.
(3) Never cut corners.
(4) Never simplify.
(5) Don't pass the buck.
(6) Never stop learning.

Some of these "commandments" may overlap with the above while others may seem to have little to do with ethics. But they crop up repeatedly in disasters and it is an undeniable fact that neglecting them has caused serious engineering failures.

9.8 Recommendations

- **Earthquake countries of Latin America should join forces in order to create durable institutions for monitoring disaster culture and keeping the memory of past disasters alive.**
  We learn from history that we never learn from history (Hegel). Should not this mean that we ought to keep trying?
- **There is scope for an international organization to field Emergency Missions after significant earthquakes in Latin America.**
  Such Missions were once sponsored and organized by UNESCO. They were integrated by Latin American experts on a voluntary basis and they were deployed within 2 or 3 days, subject to permission by the respective government.
- **More imaginative means should be found for performing cutting-edge research on megaquakes.**
  Megaquakes are rare events but they should not catch us unprepared. The opportunity of surveying a megaquake from the air within hours after it occurred should not be missed.
- **High-rise reinforced-concrete buildings should not be built on soft ground.**
  They should be supported on the hardpan underneath.
- **Interdisciplinary studies on disasters should be established at major universities in the Latin American region.**
  A modern university needs to pay attention to the urgent needs of society. It must work together with government and business. Disaster studies are badly needed in the region. Higher education can be combined with imaginative research and frontier advice to policymakers.
- **Transelastic phenomena should be studied within the context of soft matter physics.**
  The transelastic state is a region of phase space in the no-man's land between solids and liquids. Soils are granular materials that undergo a transelastic transition under intense vibration. Intergranular cohesion decays and gravity takes over as the main restoring force in wave propagation. This transition may be the cause of significant damage on soft ground.
- **Decision-making on earthquake risk in Latin America should be holistic.**
  Social science is not always a participant in important decisions which involve the earth sciences and their impact on society. For example, Luhmann (1993) has offered an extensive discussion of earthquake risk and nuclear technology which did not receive the attention it deserves.

9.8 Recommendations

The holistic approach means essentially that all points of view—that of social scientists as well as of earth scientists—should be taken into account in decisions involving earthquake risk. Social science offers new important insights that should be considered in decision-making.

Mary Douglas once pointed out that the risk of living next to a nuclear power station might be compared to that of driving an extra three miles per year. But this argument hardly impresses anyone. The reason is that the comparison is irrelevant: driving three extra miles a year is not a disaster. Such numerical arguments are often used by both sides of a major decision. We need the kind of balanced approach that social science can provide.

In conclusion we may say that an admittedly superficial re-examination of disastrous earthquakes in Latin America—the sort of treatment that can be compressed in less than 100 pages—has led us to discuss a large variety of interesting problems of theoretical and practical importance.

References

Luhmann N (1993) Risk: A sociological theory. Aldine de Gruyter, New York
Goodman RE (1999) Karl Terzaghi, Am. Soc. Civil Eng. Washington, ISBN 0-7844-0364-3, and other Terzaghi biographies

GPSR Compliance

The European Union's (EU) General Product Safety Regulation (GPSR) is a set of rules that requires consumer products to be safe and our obligations to ensure this.

If you have any concerns about our products, you can contact us on

ProductSafety@springernature.com

In case Publisher is established outside the EU, the EU authorized representative is:

Springer Nature Customer Service Center GmbH
Europaplatz 3
69115 Heidelberg, Germany